Environmental Division 2022

AA009238

Held at the 2022 AIChE Annual Meeting

Phoenix, Arizona, USA
13-18 November 2022

ISBN: 978-1-7138-7902-2

Printed from e-media with permission by:

Curran Associates, Inc.
57 Morehouse Lane
Red Hook, NY 12571

Some format issues inherent in the e-media version may also appear in this print version.

Copyright© (2022) by AIChE
All rights reserved.

Printed with permission by Curran Associates, Inc. (2023)

For permission requests, please contact AIChE
at the address below.

AIChE
120 Wall Street, FL 23
New York, NY 10005-4020

Phone: (800) 242-4363
Fax: (203) 775-5177

www.aiche.org

Additional copies of this publication are available from:

Curran Associates, Inc.
57 Morehouse Lane
Red Hook, NY 12571 USA
Phone: 845-758-0400
Fax: 845-758-2633
Email: curran@proceedings.com
Web: www.proceedings.com

TABLE OF CONTENTS

ENVIRONMENTAL DIVISION AWARDS AND HONORS (INVITED TALKS)

POSTER SESSION: ENVIRONMENTAL DIVISION

ATMOSPHERIC CHEMISTRY AND PHYSICS: MODELING AND FIELD STUDIES

ATMOSPHERIC CHEMISTRY AND PHYSICS: LABORATORY STUDIES

ADVANCED TREATMENT TECHNOLOGIES FOR WATER

REMEDIATION OF EMERGING CONTAMINANTS AND LEGACY COMPOUNDS

WATER REUSE AND RECYCLING

FUNDAMENTALS AND APPLICATIONS FOR MUNICIPAL SOLID WASTE TREATMENT AND VALORIZATION

WASTE PLASTIC - RECYCLE, REUSE AND REMEDIATION STRATEGIES

FUNDAMENTALS OF ENVIRONMENTAL KINETICS AND REACTION ENGINEERING

FUNDAMENTALS OF FOOD, ENERGY, WATER SYSTEMS

SUSTAINABILITY FUNDAMENTALS AND METRICS APPLICATIONS

FUNDAMENTALS AND APPLICATIONS FOR WASTE TREATMENT AND VALORIZATION

CO2 INDUSTRIAL, ENGINEERING AND RANDD APPROACHES

ENVIRONMENTAL ISSUES INVOLVING BIOCHAR

SUSTAINABLE FUEL FROM RENEWABLE RESOURCES

DESIGN AND ANALYSIS OF CARBON CAPTURE AND NEGATIVE EMISSIONS TECHNOLOGIES - EXPERIMENTAL

DESIGN AND ANALYSIS OF CARBON CAPTURE AND NEGATIVE EMISSIONS TECHNOLOGIES - MODELS

DESIGN AND OPTIMIZATION OF INTEGRATED ENERGY SYSTEMS

DESIGN AND ANALYSIS OF SUSTAINABLE CARBON CAPTURE AND EMISSIONS CONTROL TECHNOLOGIES

ENVIRONMENTAL ADVANCE IN NUCLEAR AND HAZARDOUS WASTE PROCESSING AND DISPOSAL

ADVANCED TREATMENT TECHNOLOGIES FOR WATER II

WASTE PLASTIC - RECYCLE, REUSE AND REMEDIATION STRATEGIES II

DESIGN AND OPTIMIZATION OF INTEGRATED ENERGY SYSTEMS II

REMEDIATION OF EMERGING CONTAMINANTS AND LEGACY COMPOUNDS II

PARTICLE MECHANICS AND SOLIDS HANDLING IN THE PROCESS INDUSTRIES

116a Lawrence K. Cecil Award Winner Presentation: 'Life-Cycle-Based Multiscale Sustainability: Challenges and Opportunities in the Era of Industry 4.0'

Yinlun Huang

The quest for sustainability reflects a crucial paradigm shift in the 21st century, which is a transition from environmental management to comprehensive sustainability assessment, followed by systematic, robust solutions for sustained economic development, environmental cleanness, and social responsibility. However, sustainability science is far from exact, and engineering sustainability is very complex in terms of scope, contents, and spatial/temporal aspects. Industrial sustainability not only studies the status quo of systems, but more importantly pathways towards a more sustainable future, all under information uncertainty and insufficient knowledge. Thus, how to characterize, analyze, design, restructure, and operate systems, from the material-product-process scale to the complex industrial system scale, in the context of life-cycle-based multiscale sustainability, is an exceedingly challenging area that needs much exploration.

In this presentation, the challenges in the research on life-cycle-based multiscale sustainability as well as the opportunities provided by Industry 4.0 technologies will be discussed. In particular, the potentials of IIoT-assisted, Digital Twin-enabled technologies will be exploited. We will show that a data and information rich environment by these technologies can greatly facilitate the realization of multi-life-cycle-stage dynamic sustainability assessment and decision making at multiple length-time scales of the system under investigation. We will introduce a preliminary framework of the life-cycle-based, digital-twin-enabled multiscale sustainability. A few case studies on sustainable surface coating/finishing will demonstrate some methodological efficacy.

116b Early Career Award Winner Presentation: 'Enabling Solvent Recovery and Reuse via Systems Engineering and Synergistic Industry-Academia Collaborations'

Kirti Yenkie

With the rapid expansion of the chemical market over the next decade, the concerns for environmental sustainability have increased due to the excessive disposal of chemical solvents. Solvents are typically used as extraction, cleaning, and purification agents and account for up to 90% of the process by mass [1]. To date, there is no proper mitigation plan to handle the inevitable rise in waste volume. These solvent wastes are typically incinerated or released after a single use. However, incineration and direct release are not green methods because of their potentially detrimental effects on the environment [2].

Thus, a systematic framework that employs Process Systems Engineering (PSE) principles, as well as data and timely inputs from Industrial Partners, has been developed to handle wastes from solvent-consuming industries. This Solvent Recovery framework considers multiple alternative recovery and purification technologies simultaneously through a superstructure-based optimization approach [3], [4]. The technologies are represented as mathematical models, which include material and energy balances, utility requirements, design, and operational constraints, as well as sustainability metrics of SPI (Sustainable Process Index) [5] and Emergy [6]. The recovery of solvents from multi-component waste streams is modeled as MINLP (Mixed-Integer NonLinear Programming) problem [7] and the competing objectives of minimizing costs and maximizing process sustainability are handled via the goal programming [8] approach. The results from our work suggest that selecting solvent recovery options require lower costs in comparison to typical incineration and reduce the emission of harmful substances into the environment. Furthermore, a user-friendly software tool that can assess solvent-containing industrial waste streams and predict the best solvent recovery pathways is also developed. This tool could be highly beneficial to pharmaceutical and chemical industries and can lead to better recycling and reuse practices.

In addition to this, a brief overview of the other ongoing projects in our Sustainable Design andamp; Systems Medicine Lab [9] including wastewater treatment, asset management, petroleum pipeline operations, and plastics recycling will be presented. The benefits of synergistic collaborations with industries and government agencies are highlighted in all our projects.

References:

[1] R. A. Sheldon, "The E factor 25 years on: the rise of green chemistry and sustainability," *Green Chem.*, vol. 19, no. 1, pp. 18–43, Jan. 2017, doi: 10.1039/C6GC02157C.

[2] C. S. Slater, M. J. Savelski, G. Hounsell, F. Urbanski, J. Geiger, and D. Knoechel, "Green Engineering Analysis of a Multi-use Solvent Recovery System for Small Volume Waste Steams in the Pharmaceutical Industry," presented at the 2010 Meeting of the American Institute of Chemical Engineers, Salt Lake City, UT, 2010.

[3] J. D. Chea, A. L. Lehr, J. P. Stengel, M. J. Savelski, C. S. Slater, and K. M. Yenkie, "Evaluation of Solvent Recovery Options for Economic Feasibility through a Superstructure-Based Optimization Framework," *Ind. Eng. Chem. Res.*, vol. 59, no. 13, pp. 5931–5944, Apr. 2020, doi: 10.1021/acs.iecr.9b06725.

[4] E. A. Aboagye, J. D. Chea, and K. M. Yenkie, "Systems level roadmap for solvent recovery and reuse in industries," *iScience*, vol. 24, no. 10, p. 103114, Oct. 2021, doi: 10.1016/j.isci.2021.103114.

[5] M. Narodoslawsky, "Chapter 3 - Sustainable process index," in *Assessing and Measuring Environmental Impact and Sustainability*, J. J. Klemeš, Ed. Oxford: Butterworth-Heinemann, 2015, pp. 73–86. doi: 10.1016/B978-0-12-799968-5.00003-8.

[6] F. Berrios, D. E. Campbell, and M. Ortiz, "Emergy-based indicators for evaluating ecosystem health: A case study of three benthic ecosystem networks influenced by coastal upwelling in northern Chile (SE Pacific coast)," *Ecol Indic*, vol. 95, pp. 379–393, Dec. 2018, doi: 10.1016/j.ecolind.2018.07.055.

[7] K. M. Yenkie, W. Wu, and C. T. Maravelias, "Synthesis and analysis of separation networks for the recovery of intracellular chemicals generated from microbial-based conversions," *Biotechnology for Biofuels*, vol. 10, p. 119, May 2017, doi: 10.1186/s13068-017-0804-2.

[8] U. Diwekar, *Introduction to applied optimization*, 2. ed. New York: Springer, 2008.

[9] "Yenkie Research Group," *Yenkie Research Group*. https://yenkiekm.com/ (accessed Jul. 15, 2022).

116c Graduate Student Paper Award, 1st Place: 'Enhancing CO Oxidation Activity via Tuning a Charge Transfer Between Gold Nanoparticles and Supports'

Haotian Yang

Charge transfer from the supports to nanoparticles at the interface is one of the key factors to determine the catalytic performances of supported nanoparticles. In this work, we showed in a systematic way that the charge transfer from semiconductor supports to Au nanoparticle catalysts can lower the onset temperature toward CO oxidation. For this study, a novel Au/SiO2/Si composite system synthesized by the helium droplet deposition method with precisely tuned SiO2 layer thickness was fabricated to control the magnitude of interfacial charge transfer. With the support of X-ray photoelectron spectroscopy and numerical simulations, it was demonstrated that the Schottky barrier formed across the Au/SiO2/Si heterojunction led to a negative charge accumulation on the surface of Au nanoparticles. In turn, this additional charge can be transferred to the antibonding orbital of adsorbed O2 molecules to activate the O–O bonds, leading to enhanced CO oxidation. In addition to the charge transfer mechanism, the role of a strong electric field arising from the formation of the Schottky barrier was also explored to explain the observed enhancement of catalytic reactivity. Overall, this work highlights an important pathway for systematically tuning metal–support interactions to accelerate catalytic reactions and designing the next generation of nanocatalysts.

Publication:

Enhancing CO Oxidation Activity via Tuning a Charge Transfer Between Gold Nanoparticles and Supports, Haotian Yang, Jiajie Cen, Qiyuan Wu, Claron J. Ridge, Xiao Tong, Chenyu Zhou, Vijayen Veerasamy, Dong Su, C. Michael Lindsay, Mingzhao Liu, and Alexander Orlov

The Journal of Physical Chemistry C 2022 126 (10), 4836-4844

DOI: 10.1021/acs.jpcc.1c10072

116d Grad Student Paper Award, 2nd Place: 'A New Avenue for Separating Contaminants from Water using Nanofiltration and Adsorptive Materials'

Francisco Leniz-Pizarro

Challenges associated with water separation technologies for per- and polyfluoroalkyl substances (PFAS) require efficient and sustainable processes supported by proper understanding of the separation mechanisms. In this study, we investigate the separation mechanism of PFASs using nanofiltration (NF) membranes and then we propose a new separation technology that aims to leverage two separation mechanisms: exclusion and adsorption. First, monovalent organic anions with similarly effective diameters (0.55-0.6 nm) were filtrated using nanofiltration (NF) membranes. The solute rejections at pH values near the membrane isoelectric point were compared to the size and mass transfer-dependent modeled rejections of these compounds in an ionized state. We find that the low pKa value of perfluorooctanoic acid (PFOA) relates to higher solute exclusions by preventing the presence of the protonated organic compound that would preferably partition into the membrane domain. The effects of Donnan exclusion are moderate and co-ion transport also contributes to the PFAS rejection rates. The contribution of an additional barrier, however, is responsible for the high PFAS rejection rates when using NF membranes. Then, we successfully synthesized a NF layer on top of a poly-N-isopropylacrylamide (PNIPAm) pore functionalized support structure. This membrane adsorbs 178 mg/m^2 membrane of PFOA at an equilibrium concentration of 70 mg/L, and rejects 70% PFOA and 50% of the smaller size heptafluorobutyric acid from solutions containing 70 μg/L of these compounds. The temperature-responsive behavior of the PNIPAm is reflected on the water permeance of the thin-film composite membranes, which varies from 7 to 24 ($L\ m^{-2}\ h^{-1}$).

Publication:

Dual-Functional Nanofiltration and Adsorptive Membranes for PFAS and Organics Separation from Water, Francisco Léniz-Pizarro, Ronald J. Vogler, Phillip Sandman, Natalie Harris, Lindell E. Ormsbee, Chunqing Liu, and Dibakar Bhattacharyya

ACS ESandamp;T Water 2022 2 (5), 863-872

DOI: 10.1021/acsestwater.2c00043

116e Grad Student Paper Award, 3rd Place: 'Effect of Combustion Particle Morphology on Biological Responses in a Co-culture of Human Lung Epithelial and Macrophage-like Cells'

Kamaljeet Kuar

Background. Freshly generated combustion particles are hydrophobic and form a fiber-like agglomerate with an average fractal dimension of 1.8. However, after aging in the atmosphere, particles become more hydrophilic allowing adsorption of water, which later exerts capillary forces between the primary particles of the agglomerate, causing the structure to collapse and become more compact spheres-like. Previous studies with fresh and aged combustion particles have linked the changing chemical composition to be the prime suspect behind the observed differences in toxicological response. However, little is known about the contribution of morphological differences in atmospherically aged particles to their toxicological response, possibly due to the difficulty in resolving the two properties (composition and morphology) that change simultaneously. A method to change particle morphology without the need to change chemical composition would help elucidate the effect of this morphological change on cellular responses.

Methods. This study altered the shape of lab-generated combustion particles, without affecting the chemical composition, from fiber-like to a more compact spherical shape using a water condensation-evaporation method. Using an electrostatic field-based air-liquid exposure (ALI) chamber, the two shapes were exposed to a co-culture of human airway epithelial (A549) and differentiated human monocyte (THP-1) cells at ALI conditions.

Results. For the same mass dose of 2 mg/cm^2, both shapes were ingested by cells, and induced a pro-inflammatory response (IL-8 and TNFα) and enhanced CYP1A1 gene expression compared to air controls. The more compact spherical particles (representative of atmospherically aged particles) induced more early apoptosis and release of TNFα compared to the more fiber-like particles.

Conclusions. The result suggests a contribution of morphology to the increased toxicity of aged particles.

Publication:

Effect of Combustion Particle Morphology on Biological Responses in a Co-culture of

Human Lung and Macrophage Cells, Kamaljeet Kaur, Raziye Mohammadpour, Hamidreza Ghandehari, Christopher A. Reilly, Robert Paine, Kerry E. Kelly

Atmospheric Environment 2022, 284

DOI: 10.1016/j.atmosenv.2022.119194

533a Selective Capture and Mineralization of CO_2/SOx Gas Using Industrial Wastewater.

Won Yong Choi, Yunsung Yoo, Dongwook Lee, Kyu Jang, Jinwon Park

Moving forward from the Paris Agreement, the UN climate change conference in Glasgow (COP26) has brought the world leaders together to produce new 'building blocks" to advance in previous enactments and to accelerate the action in lowering carbon dioxide emissions. With the awareness of importance on carbon capture technology, the Glasgow pact calls for a finance support in adapting to build resilience against the climate change. While, CCS technology is acceptable for large-scale process and commercial applications, it possesses drawbacks in cost of implementation and geological restraints. The intense energy requirement for regeneration of sorbents and potential catastrophic leakage is a challenge that CCS technology has faced. To ease the burden on, CCU process is applied for permanent sequestration of greenhouse gas with producing value-added product from the waste pollutant.

Herein, a novel CCUS process is designed by introducing seawater as a stable feedstock to capture greenhouse gases from flue gas. The alkali absorbent, sodium hydroxide obtained from the seawater is used to selectively capture CO_2 and SO_2 from the flue gas. The difference in kinetic allows the absorbent to preferentially capture the SO_2 even the partial pressure of SO_2 is low as in ppm level. The suggested process concept only requires a single absorbent to capture CO_2 and SO_2 without the need of an additional flue-gas desulfurization process by applying the two-batch type reactors. Each reactor selectively captures CO_2 and SO_2 gases until it is saturated and fed with calcium and magnesium ions to produce metal carbonates. The final products $CaSO_4$, $MgCO_3$ were analyzed with X-ray diffraction (XRD), Scanning electron microscopy (SEM) to verify the crystal shape of the final products. The present study investigated the absorption rate of CO_2/SO_2 gas within different molar ratios of absorbents at different partial pressures based on their kinetic reaction rate.

This research summarizes our research team's previous works based on the carbonation process using metal ions present in industrial wastewater. Abundance of metal ions present in industrial wastewater gives advantages in reliable feedstocks and allows sustainable carbon emission cycles. While, the suggested process only compromised with a single absorbent to capture both CO_2 and SO_2, it provides alternative routes to capture flue gas through an economically viable process. It also provides the meaning of sustainable development in which all the materials used in the process are acquired from the seawater.

533b Plastic Pyrolysis Gas Purification and Polymorph Control of Metal Carbonate Using Captured CO_2 By Deep Eutectic Solvents

Kyumin Jang, Won Yong Choi, Dongwook Lee, Yunsung Yoo, Jinwon Park

Due to the recent exponential increase production of plastic waste, the need for its treatment has become crucial. Pyrolysis technology, which pyrolyzes plastic waste to generate useful materials through it, is currently gaining a lot of attention. By definition, pyrolysis is the thermal decomposition of a material in an oxygen-free condition. In pyrolysis, the plastic waste feedstock is introduced into a reactor at 450–600 °C to produce a vapor which consists of condensable and non-condensable fractions. After condensation, the condensed vapor which is known as pyrolysis oil can be used as an auxiliary fuel and a chemical raw material. However, non-condensing vapor is a mixture of gases containing large amounts of CO_2, H_2, CH_4, and C_2H_4 approximately 15%, 49%, 22%, and 14%, respectively. It is generally used as a fuel for pyrolysis that produces heat. Non-condensing gas contains about 15% of CO_2, it reduces calorific value and utilization. To increase the calorific value and efficiency of utilization, it is necessary to remove CO_2 from non-condensing gas. To do so, the carbon capture, and utilization (CCU) technology that presents several available processes for use can be applied. Mineral carbonation, among CCU technologies, can sequester CO_2 on a large scale; does not require additional energy consumption because no catalyst is required for metal carbonate formation, and the generated carbonate can be reused or sold, finally improving the economy of the process.

Deep eutectic solvents (DESs), eutectic mixtures of two or more components by hydrogen bond, have been identified as a novel class of Ionic liquids(ILs). DESs are synthesized by mixing a hydrogen bond acceptor (HBA) and a hydrogen bond donor (HBD) and have similar properties and characteristics as ILs, such as existing as a liquid below 100°C. However, DESs have a high viscosity disadvantage via hydrogen bonding, which causes the gas mass transfer problem. To enhance the gas mass transfer of DESs, H_2O is added to compensate for the high viscosity of DESs. H_2O can serve as a lubricant with very low friction while in liquid absorption using DESs. Furthermore, the ILs-H_2O system has also been used to create a shape-specific and shape-controlled synthesis of nanoparticles.

Throughout this study, DESs were synthesized by mixing CO_2-active site-containing HBD such as monoethanolamine and used HBA such as choline chloride, and tetrabutylammonium bromide. Then to compensate for the high viscosity of DESs, H_2O was added and CO_2 contained in the non-condensing gas was removed by the synthesized DESs-H_2O system. Following that, mineral carbonation, among CCU technologies, was applied to utilize the absorbed CO_2 without additional energy. In the mineral carbonation process, the CO_2 in CO_2-saturated DESs reacted with calcium ions, and then the absorbed CO_2 and calcium ions are converted into calcium carbonate. Furthermore, the Fourier transform infrared spectroscopy (FT-IR) and ^{1}H nuclear magnetic resonance (NMR) analyses were conducted for the investigated samples to analyze the absorption mechanism between CO_2 and DESs, and to compare changes in the DES properties depending on the water contents. In addition, the crystallographic properties of the calcium carbonate were investigated using X-ray diffraction (XRD) analysis. FT-IR analysis was used to validate the non-crystallographic properties of the calcium carbonate which is are not detected via XRD analysis. Finally, scanning electron microscopy (SEM) was used to verify the crystal shapes of the product and support XRD and FT-IR results.

533c Perstraction: A Membrane-Assisted Liquid-Liquid Extraction Process for Removing Pfas from Water

Catherine Almquist

We are developing and characterizing a novel perstraction process for the removal of perfluorinated compounds (PFAS) from water. The process utilizes a membrane barrier between an organic solvent and a PFAS-contaminated water stream. The membrane barrier is required in this process, since significant foaming or emulsion formation occurs at the interface between organic solvents and PFAS-contaminated water. For this project, we are synthesizing membranes using polydimethylsiloxane (PDMS), which is a hydrophobic material. The inclusion of nanoparticles within the PDMS can vary the properties of the membrane, and so we have investigated the effects of various nanoparticles imbedded in PDMS membranes on the absorption of selected solvents and perfluorooctanoic acid (PFOA). We have found that nanoparticles imbedded into the PDMS membrane significantly increase the absorption of PFOA, a selected representative of PFAS compounds.

The objectives of this project are to design, construct, and assess the performance of a prototype perstraction system. Ancillary objectives are to determine the partition coefficients of PFOA between water and selected solvents; determine the solubility and diffusivity of the selected solvents in PDMS membranes; and to assess the effects of selected nanoparticles on the solubility and diffusivity of selected solvents in PDMS and on the uptake of PFOA in PDMS (both extent of PFOA uptake and kinetics of the uptake).

533d A Study on Imidazole-Alkanolamine-Based Deep Eutectic Solvents (DES) Functionalized for Low-Concentration Sulfur Dioxide Absorption

Dongwook Lee, Yunsung Yoo, Kyu Jang, Won Yong Choi, Jinwon Park

SO_2 emitted in the process of using fossil fuels adversely affects the human body and air quality, so emissions are strictly controlled by related facilities. The biggest contribution to this is SO_2 removal technology called flue gas desulfurization (FGD).

In the FGD process, SO_2 is removed by spraying limestone slurry, which is widely used due to its technical stability and simplicity. However, the FGD process can potentially cause environmental problems by producing waste and wastewater. Also, in countries with narrow territories such as Republic of Korea, there is a problem with the supply of raw materials because of limited reserves of high-grade limestone. Therefore, we studied wet absorption methods that are more universal and can be used as absorption as candidates for alternative technologies.

Among many absorbents, the use of Deep eutectic solvent (DES) with good tunability and stability as the SO_2 absorbent is considered in this study. DES is prepared by mixing hydrogen bond donors (HBD) and hydrogen bond acceptors (HBA). DES has low vapor pressure and high thermal stability due to hydrogen bonding, making it suitable as an absorbent. In addition, since various materials capable of hydrogen bonding can be selected as precursors, it can be functionalized to be advantageous for SO_2 absorption.

As an acid gas absorbent, one of important factor is the absorption performance at low concentration. According to Henry's law, high concentrations of gas are well absorbed, and low concentrations of gas are poorly absorbed. Since there are facilities where only low concentrations of SO_2 are emitted, it is necessary to use an absorbent that can chemically react with SO_2.

In this study, DES, which has stable physical properties and can efficiently absorb and reabsorb low concentrations of SO_2, was studied. As the HBA of DES, imidazole expected to have excellent reactivity to SO_2 was used. Another focus of this study is the use of SO_2-philic HBDs. In other DES absorbent studies, since research on HBAs is a major issue, substances that do not react with acidic gases such as glycerol and ethylene glycol are used as HBD. However, in this study, the absorption capacity per mass of DES was maximized by using alkanolamine, which is expected to absorb SO_2 as HBD.

The thermal stability changed with the formation of the proposed DES was measured. In addition, a gas absorption experiments were performed to measure the absorption capacity of low-concentration SO_2, and absorption and desorption experiments were repeated to analyze the reusability of the absorbent. As a result, the thermal stability of all precursors used increased as DES was formed. At a concentration of 500 ppm SO_2, the absorption capacity of proposed DES is more than double that reported in related studies. In the reabsorption experiments, the negligible performance decrease was observed due to the loss of the absorbent. At last, the SO_2 absorption mechanism was analyzed through ^{1}H NMR analysis.

533e Thermo-Responsive Hydrogel with Deep Eutectic Mixture Co-Monomer As Drawing Agent for Forward Osmosis

Anelyn Bendoy, John Edward Sio, Hana Zeweldi, Hern Kim, Wook-Jin Chung, Grace Nisola

Forward osmosis (FO) is an emerging alternative technology to membrane-based separation process for wastewater treatment and seawater desalination such as reverse osmosis (RO). Unlike RO that requires high pressure in producing clean water, FO employs minimal osmotic pressure difference across the FO membrane. Thus, membrane fouling is less likely to occur in FO which contributes to its energy-efficiency. Recently, membrane development and regenerable draw solutions (DS) are the two areas of interests in various researches for FO technology to progress industrially. The challenge in the DS development is the production of a recyclable DS that will reduce the energy expenditure in post-FO separation of DS and clean water production. Attempts in addressing such challenge include the use of stimuli-responsive DS such as magnetic nanoparticles and hydrogels. However, even if magnetic nanoparticles have high osmotic pressure, the issues on agglomeration and membrane fouling are still a challenge. While in hydrogels, enhancement of the inherent osmotic pressure as well as the release of absorbed water are still under diligent exploration.

Herein, N-Hexyl-N,N-dihydroxyethyl-N-methylammonium chloride–acrylic acid ([DHEA]Cl-AA) deep eutectic mixture (DEM) was used as a co-monomer with N-isopropylacrylamide (NIPAM) to form a thermo-responsive hydrogel. The purpose of which is to enhance the osmotic pressure of the resulting hydrogel without compromising its thermo-responsive property when used as a drawing agent for FO. FT-IR, H^1 NMR and TGA were used to confirm the structure of the synthesized DEM. Meanwhile, FT-IR, DSC and SEM analysis were used to characterize the synthesized hydrogels. Swelling and de-swelling properties of the hydrogels were evaluated by soaking the hydrogel in DI water for 72 h and heating at 45 °C for 30 min, respectively. The water absorbing property of the hydrogels was also investigated as a drawing agent against 2000 ppm NaCl feed solution in FO system. Furthermore, water recovery rate of the swollen hydrogels after FO run was also examined by the de-swelling method mentioned above.

Results reveal an overall improvement in the water absorbing capacity of P(NIPAM-*co*-DEM) hydrogels relative to the pure PNIPAM hydrogel in terms of swelling ratio and FO water flux. De-swelling data showed an enhanced water release when DEM was co-polymerized into the NIPAM backbone. This study demonstrates the potential of DEM as a co-monomer in producing a thermos-responsive hydrogel with improved FO water flux and water release ability for DS regeneration.

This research was supported by NRF funded by Basic Science Research Program through the Ministry of Education (2021R1I1A1A01050003 and 2020R1A6A1A03038817).

533g Surface Phenomenon Affecting Removal Efficiency of Nitrate from Water on Dispersed Single Atoms in Cu Metal Catalyst: An Ab-Initio Study

Srishti Gupta, Adam Chismar, Matthew Shaffer, Daniel J. Rivera, Christopher Muhich

Anthropogenic activities have disturbed the balance of the delicate nitrogen cycle. Overuse of nitrate (NO_3^-) rich fertilizers have led to the leaching of nitrate in our water system. If nitrogen-rich water is consumed by humans at concentrations higher than 10 mg/L, it can cause methemoglobinemia, or "blue baby syndrome", thyroid disease, fatigue, weight gain, hair loss, goiters and colon cancer. Therefore, Nitrate remediation is an important step to re-balance the nitrogen cycle. Thermochemical and Electrocatalytic nitrate reduction is a potential remediation process for rebuilding a circular economy by reducing NO_3^- to NH_3 or N_2. However, both are currently not commercially viable because of low selectivity towards safe (N_2) or reusable (NH_3) products. Nitrate reduction reaction often competes with other electrochemical and thermochemical reactions such as hydrogen evolution reaction (HER). Moreover, the high cost of the best catalyst (like Pt, Pd, In etc.) makes nitrate remediation an expensive process. One method to improve the activity and selectivity of Nitrate reduction and decrease the cost of raw material is to opt for cheaper metal (e.g., Cu, \$0.43 per mol) catalyst impregnated with Single-Atom Alloy (SAA).

Here we investigate the mechanism of electrochemical and thermochemical nitrate reduction using Density Functional theory over 6 different transition single atom alloys dispersed in Cu[111] to understand the critical effects of dispersed atoms on the activity and selectivity of the electrode surface. The potential and pH effects are included by considering the Chemical Hydrogen electrode and protonation energies of aqueous species. We find that the energy required to dissociate NO_3 Ru and Pd-SAA by thermochemical Nitrate Reduction pathway is higher than desorption on NO_3^-. The difference in energies is reversed by increasing the potential and pH of the solution in the electrochemical system. The selectivity is dependent upon the energy difference between N^* hopping from the SAA site to Cu sites and NH^* formation. For Ru-SAA, the N^* is localized near a single atom, which results in highly selective NH_3 formation. For Pd-SAA, the N^* hopping has a lower activation barrier, which leads to the final reaction occurring on Cu sites, which is dependent upon reaction potential (selective towards N_2 at potential < -0.2 V vs. HER). We use the energy of single species (H^*, O^* and N^*) as descriptors for building volcano plots to predict an ideal single-atom catalyst that can be used for highly selective NO_3RR near-neutral pH at a low cost.

533h Bench-Scale Testing of Electrochemical Recovery of Phosphorus from Post-Digester Municipal Wastewater Driven By Magnesium Salt

Lawrence Ajayi, Sana Heydarian, Jason Trembly, Damilola Daramola

Phosphorus is an irreplaceable macronutrient necessary for plant growth and other industrial processes, whereas when in excess in water bodies can lead to eutrophication which causes algal blooms, reduced water clarity, and toxicity to aquatic life due to lack of oxygen. Through human excrements, dishwashing agents, and other unspecific sources, approximately 1.8 g phosphorus per capita per day reaches the wastewater treatment plants [1]. This wastewater, low in phosphorus concentration, passes through a primary clarifier and then a digestion process. The typical digestion process results in a phosphorus-rich supernatant in the digester effluent. In addition, as demand for fertilizers increases globally and the supply of non-renewable rock phosphate steadily decreases, the costs of phosphate-based fertilizers will increase, significantly impacting the affordability of food products [2, 3]. Recovery of phosphorus from waste, as struvite ($MgNH_4PO_4 \cdot 6H_2O$): a slow-release fertilizer, is a potential pathway to solving these problems. Therefore, it is imperative to develop a technology to optimize phosphorus recovery from post-digester wastewater as this could lead towards environmental sustainability and create economic value.

This research study focuses on recovering phosphorus as struvite (or a mixed-phase solid) from post-digester municipal wastewater using an electrochemical technology. The electrochemical technique adopted provides localized pH control which is necessary for struvite formation through oxygen reduction at the cathode. This approach also aims to achieve a process that enables easy automation with little or no human intervention. Belarbi *et al.* investigated the impact of the presence of other dissolved species specifically Ca^{2+} in wastewater and how it can alter the chemistry of struvite precipitation [4]. This provided insight into negating the effect of this competing ion by increasing the concentration of magnesium in the wastewater in addition to other major ions in this stream including Na^+, K^+, SO_4^{2-} and Cl^-. In a bid to potentially recover phosphorus as struvite in this study, magnesium will be supplemented by the introduction of magnesium salt.

The effect of the available surface area of the working electrode, flow turbulence, applied potential, and dissolved oxygen present will be investigated. Also, the percentage recovery of phosphorus as struvite will be determined while analyzing it using various characterization techniques such as scanning electron microscope, Energy-dispersive X-ray spectroscopy, Raman spectroscopy, and X-Ray Diffraction.

References:

1. Cornel, P., andamp; Schaum, C., "Phosphorus recovery from wastewater: needs, technologies and costs" *Water Science and Technology*, 59(6), 1069–1076, 2009.
2. László Kékedy-Nagy, Leah English, Zahra Anari, Mojtaba Abolhassani, Bruno G. Pollet, Jennie Popp, Lauren F. Greenlee, "Electrochemical nutrient removal from natural wastewater sources and its impact on water quality" *Water Research,* 210, 2022.
3. Arseto Yekti Bagastyo, Anita Dwi Anggrainy, Khoiruddin Khoiruddin, Riang Ursada, IDAA Warmadewanthi, I Gede Wenten, "Electrochemically-driven struvite recovery: Prospect and challenges for the application of magnesium sacrificial anode" *Separation and Purification Technology*, vol. 288, 2022.
4. Z. Belarbi, D. Daramola and J. P. Trembly, "*Bench-Scale Demonstration and Thermodynamic Simulations of Electrochemical Nutrient Reduction in Wastewater via Recovery as Struvite*" J. Electrochem. Soc., vol. 167, no. 4, Nov. 2020.

533i A Life Cycle Assessment Study of Wind Power Based Multi- Source Energy Systems

Joon Heon Lee, Jun Hyung Ryu

Environmental impact has become one of the most important issue in developing new energy systems under the strengthening international regulations due to increasing financial costs caused by climate change.

The emission amount of carbon and other greenhouse causing materials should be carefully calculated to mitigate the impact of global warming. Due to global disasters caused by climate change, it has become clear that there should be no exemptions in the effort to reduce the emission of greenhouse gases.

Rigorous analysis should be conducted covering the entire life span of the corresponding energy system including the site preparation, construction, operation and decommission to assess their environmental carbon footprint. Besides, parts and materials should be closely monitored including their manufacturing and associated supply chains.

Recently many wind power generation facilities have been constructed as a non-fossil fuel energy source. Obviously there are many advantages in using the wind power: no raw materials should be purchased once it has been set up, no emission of environmentally hazardous gases, etc. With the decreasing power generation cost by using wind power, the size of wind power generators is increasing. The off shore wind turbines are actively built around the world. At the same time, attention has been given to take the best utilization of redundant energy from renewable energy sources including wind powers. One promising alternative is to transform the redundant electricity into hydrogen and use it as transportation energy source.

Some countries including South Korea, Japan, etc. are actively preparing hydrogen based transportation energy system as an alternative to gasoline based ones. This work is motivated by the need that such an effort should be carefully reviewed to prevent the unexpected troubles that are often encountered.

This paper investigate the overall process of site preparation, construction, operation and decommissioning of wind power turbine, electrolysis and hydrogen refueling station. Land and sea transport were considered for some parts from the manufacturing sites to the construction and setup sites. It was inevitable to make some assumptions because it is not possible to gather up-to-date information on some of them. The computation has been made by replacing them as globally used alternatives.

A case study addressing an energy system including on-shore wind power generation farms in the Gyeongbuk region of South Korea is presented in this paper. The case study of Life Cycle Assessment (LCA) of the energy system has been conducted by using widely used commercial LCA software of SimaPro. Lessons learned during the field study have been discussed as well as limitations of the current case study.

Acknowledgement

This work was supported by Korea Institute of Energy Technology Evaluation and Planning (KETEP) grant funded by the Korean government (MOTIE) (No. 20214000000010, NRF-2020R1I1A3A04038008).

533j The Characterization of Corrosion Inhibitors from Quinazolinones and Benzoxazines Derivatives By Electrochemical Methods

Muzammil Nishar Ahmed, Nishal Egodawaththa, Noel Manring, Nasri Nesnas, Pavithra Pathirathna

Corrosion of metals remains one of the most challenging aspects of industrial maintenance as it causes gradual damage to various vital parts and instruments of an industrial process. It is also one of the major setbacks of many marine systems as they are continually exposed to conditions ideal for corrosion. Many industries have incurred huge costs due to corrosion-related damages; therefore, the development of efficient corrosion inhibitors is of utmost urgency.

Organic compounds are among the most common corrosion inhibitors which protect the metal by adsorbing onto its surface. Earlier research has identified quinazoline compounds as corrosion inhibitors. In this study, the corrosion inhibitive properties of the following quinazolinone compounds: 3-amino-2-methylquinazolin-4(3H)-one (Q1), 3-amino-2-phenylquinazolin-4(3H)-one (Q2), and benzoxazine compounds: 2-methyl-4H-3,1-benzoxazin-4-one (B1), and 2-phenyl-4H-3,1-benzoxazin-4-one (B2) on mild steel were characterized. The structure-activity relationship (SAR) of selected derivatives shows higher adsorption to the metal surface. The corrosion inhibition depends on factors such as concentration of the organic ligand, stability of the ligand, the position of the heteroatom, and other physical factors. They were characterized by potentiometry, weight loss, and EIS experiments, which show great agreement with the hypothesis that these ligands can be used as an effective corrosion inhibitor. The selected quinazolinones and benzoxazines are much more stable across a broad range of pH, which could be attributed to the organic ligands easily adsorbing to the metal surface. Q2 and B2 have comparatively higher steric factors and therefore have a higher potential to prevent the oxidation of the metal surface. The functional groups of the selected ligands play a vital role in the H-bond formation of each ligand making a uniform monolayer. The carbonyl and amine groups of the organic ligands affect the unique adsorptive layer on the metal surface. Scanning tunneling microscopy images were also acquired to examine the surface morphology of the system which agrees with the prior experimental results. We conclude that the selected ligands can be utilized as effective corrosion inhibitors in various industries as they can withstand broad ranges of pH making them more durable.

Graphical Abstract

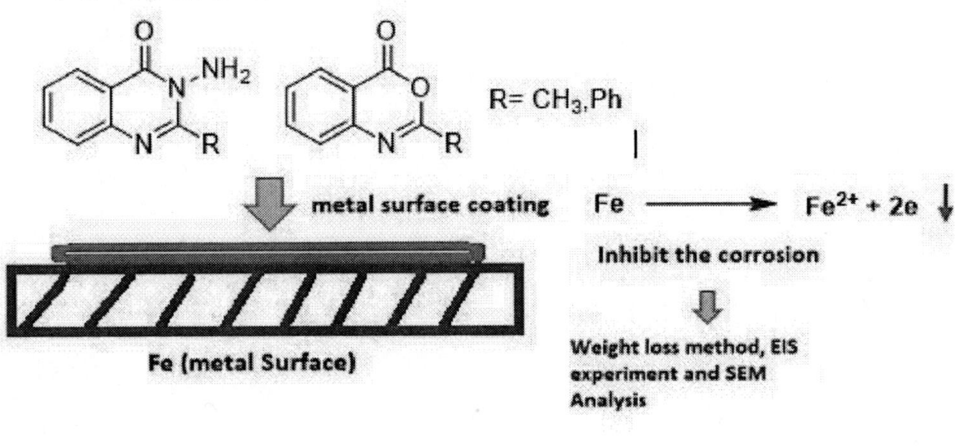

533k Understanding Integrated Carbon Negative Systems

Elizabeth Abraham, Dhabia Al-Mohannadi

With rising concerns associated with climate change, several technologies and approaches have been developed to limit the release of carbon dioxide into the atmosphere. The definite objective of these developments is to create carbon negativity as the previously released carbon dioxide must now also be dealt with in addition to the newly expected emissions created to mitigate the adverse effects of the global climate crisis. One such group of technologies that actively removes existing carbon dioxide from the atmosphere is negative emissions technologies, or NETs. Some common examples of NETs include direct air capture, bioenergy carbon capture and storage, afforestation, and enhanced weathering. NETs remove atmospheric carbon dioxide through various biological, chemical, and physical mechanisms and transfer it typically to other environmental storage compartments. Though typically deployed individually, these technologies show potential synergistic interactions with themselves and other carbon capture storage and utilization or CCUS systems. Such exchanges can be facilitated by their mutual intent to collaborate and their spatial proximity, or in other words, through industrial symbiosis. Recently, there has been an influx of tools developed to create integrated systems that effectively integrate multiple resources and processes. They determine the optimum combination of these options that leads to sustainability. The options available to them combine a variety of carbon reduction pathways, such as NETs, CCUS, and renewable energy. While it is undoubtedly possible to achieve carbon negativity with systems that incorporate NETs, systems that integrate only CCUS and renewable energy options without any NETs can also be carbon-negative. One fundamental aspect that makes this possible is clustering, where related industries are set up together to closely manage their emissions through careful exchanges. This work aims to understand how systems that are overall carbon negative with and without NETs differ from one another and, further, the role of clustering in the context of creating carbon-negative systems.

533l Magnetic Hierarchical Titanium-Ferrocyanide for the Removal of Radioactive Cs from Water

Hee-Man Yang

The selective removal of ^{137}Cs from contaminated seawater or radioactive liquid waste remains a challenge due to the presence of high concentrations of competing ions, such as K^+, Na^+, Mg^{2+}, and Ca^{2+}. In this study, the Magnetic hierarchical titanium-ferrocyanide were fabricated for the removal of radioactive cesium from water with enhanced Cs adsorption and magnetic separation properties. The new magnetic Cs adsorbent has a core–shell structure that comprises a Fe_3O_4 core, an interlayer ofSiO_2, and a titanium-ferrocyanide-shell with hierarchical nanostructure. At first, the magnetic Fe_3O_4 nanoparticles synthesized via a hydrothermal reaction were coated with SiO_2. Then, TiO_2 were coated on the surface of SiO_2 coated magnetic nanoparticles. Finally, the hierarchical titanium-ferrocyanide composed of 2-dimensional TiFC flakes was fabricated on the surface of core-shell Fe_3O_4@SiO_2@TiO_2 microparticles using a TiO_2 sacrificial template via a simple reaction with potassium ferrocyanide (FC) based on the Kirkendall-type diffusion.

The resulting magnetic Cs adsorbent shows much faster Cs adsorption kinetics and higher adsorption capacity than other magnetic Cs adsorbents because of the increased effective surface area of hierarchical titanium-ferrocyanide. Detailed Cs adsorption studies revealed that the adsorption isotherms and kinetics could be effectively described by the Langmuir isotherm model and the pseudo-second-order model, respectively. Moreover, the Magnetic hierarchical titanium-ferrocyanide selectively adsorbed Cs even in the presence of various competing ions, such as K^+, Na^+, Mg^{2+}, and Ca^{2+}

In radioactive tests, even the low concentration of hf-TiFC (0.1 g/L) exhibited excellent removal performances in simulated groundwater and seawater with a high removal efficiency exceeding 99.85% at an initial ^{137}Cs specific activity of approximately 70 Bq/g, which correspond to approximately 21.88 ppt. Therefore, our Magnetic hierarchical titanium-ferrocyanide has excellent potential for the treatment of various ^{137}Cs-contaminated sources.

Decrease of biological contamination from the municipal wastewater by trapping nZVI in Cellulose Acetate (CA) polymer: operating conditions, reduction mechanisms with AI applications

Maha M. El-Shafei[1], Ahmed S. Mahmoud[2,3], Mohamed K. Mostafa[4], Robert W. Peters[5]

[1]Sanitary and Environmental Institute (SEI), Housing and Building National Research Center (HBRC), Giza, Egypt.

[2]Institute of Environmental studies, Arish University, North Sinai, Egypt

[3]Scientific Research Development Unit, Egyptian Russian University (ERU), Badr, Egypt

[4]Faculty of Engineering and Technology, Badr University in Cairo (BUC), Cairo, Egypt

[5]Department of Civil, Construction and Environmental Engineering, University of Alabama at Birmingham, Birmingham, AL, USA

Abstract

Nanotechnology, particularly Zero Valent metals, is a cutting-edge technique for the adsorption and degradation of a variety of wastewater pollutants. This study prepared and characterized nano Zero Valent Iron (nZVI) to be entrapped in natural Cellulose Acetate (CA) polymer due to their efficacy, economics, and safety. Variations in BOD levels were used to assess the elimination of biological pollutants. At various pH, dosage (g/L), contact duration (min), stirring rate (rpm), and beginning concentrations, the influence of operational parameters was investigated. Adsorption isotherm and kinetic studies were also carried out in order to determine equilibrated reaction mechanisms. The Response Surface Methodology (RSM) relationships between variables and removal percentages were investigated using linear regression analysis. The importance detections of Artificial Intelligence Neural Networks (ANNs) were built using a nonlinear feed-forward backpropagation system. Finally, this research found that effective BOD removal percentages reached 76.8%. At pH 7, using a wet dose of 3g/L, 25 minutes, and 200 rpm, the maximum removal efficiency for an initial BOD concentration of 300 mg/L was observed. CA/nZVI preferred the khan isotherm model and Pseudo Second Order (P.S.O) with the lowest summation of errors 0.0934 and 0.132, respectively, according to the adsorption results and kinetic analysis. With an error range of (-2.5, +7.5 percent), the RSM and ANN results successfully estimated the removal efficiency.

1. Introduction

Water scarcity is one of the most pressing issues confronting the globe today, prompting governments to seek innovative solutions and technology to alleviate the world's water scarcity [1-3]. Desalination, wastewater treatment, and rainfall collection are examples of these solutions [4-7]. Furthermore, clean and safe water is one of the most basic human necessities required for survival and social development [1, 8-10]. Water treatment is seen as a long-term and short-term sustainable solution to water constraint. Because of the lack of soil filtering, many biological pollutants such as bacteria, viruses, protozoa, and other transportable organisms might impact the quality of wastewater [11, 12]. The fecal coliform group of organisms is the most well-known, followed by the fecal streptococci bacteria [13-15]. The presence of biological pollutants is a key sign of sewage contamination [16-18]. One of the most worrying protozoa is "Giardia lamblia," which is found in many surface waters and is discharged as a cyst in animal feces [19-22]. Stable cysts are carried to the subsurface by sinking streams [23, 24].

Wastewater treatment has several advantages since it is not only a source of clean reuse water, but it also has the ability to create energy through methane, minimize trash creation in the environment, and provide natural fertilizers [25, 26]. Wastewater is defined as dirty water used by commercial, industrial, and residential users that contains suspended and dissolved organic and inorganic components such as lipids, carbohydrates, detergents, and microorganisms such as bacteria, algae, and nematodes [3, 27, 28]. In this research, we are interested in the spatial biological contaminant's removals of municipal wastewater treatment by measuring the quantity of water oxygen reduced by biological behavior using the BOD standard test technique.

The primary source of biological contaminants in municipal wastewater is domestic usage [29, 30]. These pollutants may be removed using a variety of ways such as physical, chemical, and biological processes; however, the physical treatment procedure generates highly concentrated sludge that is extremely poisonous and difficult to remove [22, 31-33]. Also, for the biological process, it is frequently used, although it is sluggish and has restrictions owing to the presence of non-biodegradable contaminates, which causes toxicity to microorganisms at times [6, 34-36], To do this, we must develop more powerful and efficient municipal wastewater treatment technology [37-39]. Nanotechnology is one of the most interesting and recently discovered approaches, with incredible promise for wastewater treatment and other environmental concerns [40-42]. Nanotechnology may be classified depending on the nature of the nano-materials, with three major categories: nano-adsorbents, nano-catalysts, and nano-membranes [43]. Nano zero-valent metals (nZVI) are one of the core materials that provide outstanding results in removing inorganic and organic pollutants present in wastewater [44].

Biochemical oxygen demand (BOD) is an important metric for assessing water quality. BOD is defined as the quantity of oxygen consumed by microorganisms such as bacteria when decomposing organic matter under aerobic (in the presence of oxygen) conditions at a specific temperature; it is also used to detect organic contaminants in wastewater [45]. BOD is one of the most important elements in wastewater treatment. Although bacteria are necessary for the breakdown of organic polluted substances, too much BOD can result in undesired numbers of bacteria and a drop in oxygen levels, which can lead to the death of certain living creatures in water [46]. Because BOD removal is critical in wastewater treatment, we concentrated our research on biological pollutants removal by researching BOD changes during the treatment process under various situations [47-49].

The most widely utilized polymers in the manufacture of membranes for water desalination, wastewater purification, and biomedical applications are cellulose and its derivatives [50, 51]. This is a result of cellulose's extensive use, superior mechanical strength, thermal stability, and ease of supply. Various research has recently employed cellulose and its derivatives to entrap various nanomaterials [52-54]. Cellulose derivatives affect the characteristics of polymers and increase the capacity for removing various impurities [55]. To remove dangerous chemicals including Cr (VI),

Ni, and Cd ions from aqueous solutions, several types of composites containing hydroxyapatite were created [56-58]. Additionally, as with chitosan, poly-L-lactic acid, or gelatin, these composites show strong biological activity [59, 60]. The production of materials for osseointegration was the main focus of the nanocomposites based on cellulose derivatives and hydroxyapatite [61]. Due to their solubility in a variety of polar chemical solvents, cellulose acetate and carboxymethyl cellulose are the most often utilized cellulose derivatives [62, 63]. In the treatment of water and wastewater, trapping nZVI in cellulose acetate (CA) membrane offers various benefits, including resolving nZVI separation issues and expanding CA's pore size to lower treatment pressure. The chemical make-up of cellulose acetate is shown in Figure (1).

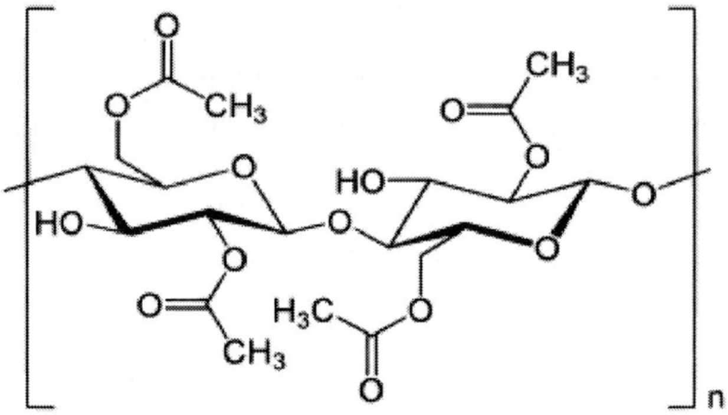

Figure 1. Structural characteristics of cellulose acetate (CA).

This study looked at employing entrapped CA/nZVI to remove biological pollutants from wastewater. The impact of operating conditions was investigated in order to increase removal efficiency. In order to comprehend the removal mechanism better, adsorption isotherm and kinetic studies were also carried out. Finally, to manage removal performance, artificial intelligence models were created.

2. Materials and methods

2.1.Preparation of nZVI

Approximately 0.0037M from $FeCl_3.6H_2O$ was liquified into 60 mL of 4 ethanol/1 deionized water mixture. The falling $NaBH_4$ solution was prepared by liquifying about 0.7564 g into 200 mL DW. According to equation 1, a burette was used to drip the reducing $NaBH_4$ solution into a ready iron solution in a drop-by-drop fashion. Direct precipitation of the dark nZVI material occurred. The prepared nZVI was rinsed with double-distilled water and dried for five hours at 75°C.

$$2FeCl_{3(aq)}+6NaBH_{4(aq)}+18H_2O_{(l)} \longrightarrow 2Fe^0_{(s)}+21H_{2(g)}+6B(OH)_{3(aq)}+6NaCl_{(aq)} \quad (1)$$

2.2.Preparation of encapsulated nZVI in Cellulose Acetate (CA/nZVI)

The CA/nZVI membrane was made using the phase inversion technique. At room temperature, 4g of cellulose acetate was dissolved in 25 ml of dimethylformamide (DMF) solution, and the dissolution was then completed by mixing the liquid at 300 RPM for 60 minutes. About 0.4 g of nZVI was added to the dissolved CA solution, stirred, and then allowed to stand for 10 minutes before being dropped into a gelation bath. It was created and heated to 17°C a 2L non-solvent distilled water gelation bath containing 0.2 weight percent sodium lauryl sulphate (SLS) and 2 weight percent DMF. The CA/nZVI beds were gathered, and then they were cleaned with distilled water. The beads in the image are shown to be about 3–4 mm in size.

2.3.Batch studies

Throughout the length of the approach, the adsorption and degradation of BOD onto CA/nZVI were examined using a batch technique and a single variable. Based on earlier research [49, 64-66], the operating conditions and their ideal ranges were chosen. The batch experiments at various operational variables, including pH, CA/nZVI dosage (g/L), contact duration (min), rate (rpm), and beginning BOD concentrations (mg/L), are reviewed in Table 1. In Erlenmeyer flasks, 300 mL of BOD solutions were equilibrated with a known weight of wet CA/nZVI approximately 3.0 g before being shaken for a predetermined amount of time at room temperature. The residual total BOD concentrations were determined in accordance with the 23rd edition of the Standard Methods for the Examination of Water and Wastewater after the equilibrium state [67]. Using Eq. 2 [68-70], the proportion of BOD removals was computed. Eq. 3 [9, 33, 71] the proportion of BOD removals was computed [72, 73].

$$Sorption\ (\%) = \left(\frac{C_0 - C_e}{C_0}\right) \times 100 \quad (2)$$

where C_o (mg/L) is the initial BOD concentrations and C_e (mg/L) is the equilibrium BOD concentrations in solution.

$$Q_e(mg/g) = \frac{(C_0 - C_e)V}{m} \quad (3)$$

where q_e (mg/g) is the equilibrium adsorption capacity, V (L) is the volume of aqueous solution, and m (mg) is the wet weight of the adsorbent [74].

Table 1. Batch tests for BOD removals under various operating conditions.

experiment No.	Dose (g)	Contact time (min)	pH	Stirring rate (rpm)	Concentration (mg/L)
Effect of pH	3	25	3-11	200	300
Effect of CA/nZVI dose	1-5	25	7	200	300
Effect of time	3	5-60	7	200	300
Effect of stirring rate	3	25	7	100-500	300
Effect of BOD concentration	3	25	7	200	100-500

2.4.Isotherm studies

With the use of nonlinear adsorption models including Langmuir, Freundlich, Toth, Hill, Sips, Redlich-Peterson, Khan, Koble-Corrigan, and Jovanovich, several isotherm investigations for

BOD elimination using CA/nZVI were conducted. These models' nonlinear equations were derived from [75].

2.5.Kinetic studies

The standard BOD solution was in contact with CA/nZVI at various intervals while receiving the same dosage and temperature in order to measure the time it took for the equilibrium state to be reached. Using Eq. 4, the quantity taken out at time t, qt (mg/g), was computed.

$$Q_t = \frac{(C_o - C_t)V}{W} \qquad (4)$$

where C_o (mg/L) is the initial BOD concentration, C_t (mg/L) is the concentration at t time, V (L) is the volume of the solution, and W (g) is the CA/nZVI dosage [76, 77].

Nonlinear pseudo-first and second-order, intraparticle, Avrami, and Elovich approaches were used to analyze the reaction process.

2.6.Validation of adsorption isotherms and kinetics

To determine whether nonlinear isotherm and kinetic models fit data most accurately and to identify the model that best captures the course of therapy, a number of nonlinear error functions were evaluated [78].

2.7.Response surface methodology

According to general Equation 5, a linear regression analysis was used to determine the relationship between removal efficiency and operational variables (pH, CA/nZVI dosage, time, and rate) on BOD removal efficiencies. The model's accuracy was assessed using the statistical coefficient of determination (r^2 value). By examining the p values, the t-test was also employed to determine the statistical significance [6, 34, 35].

$$Y = \beta_0 + \beta_1 x_1 + \beta_2 x_2 + \beta_3 x_3 + \beta_4 x_4 + \beta_5 x_5 \qquad (5)$$

where $Y_{(BOD)}$ is the predicted response of removal efficiency (%).

2.8.Artificial intelligence Neural Networks (ANNs)

To predict the individual relationships between the "total BOD" (CBOD and NBOD) and "removal efficiency" (output layer), calculated hidden layer, and operating conditions pH, dose, time, and stirring rate (input layer), two ANN models were developed and calculated using the Multilayer Perceptron Statistical Model (MLP), as shown in Fig. 2. Utilizing the activation function "Hyperbolic tangent" without omitting any outcomes, accumulation of square error, detection of the difference between predicted and actual values, and detection of the relevance and normalised importance of each operational variable [79, 80].

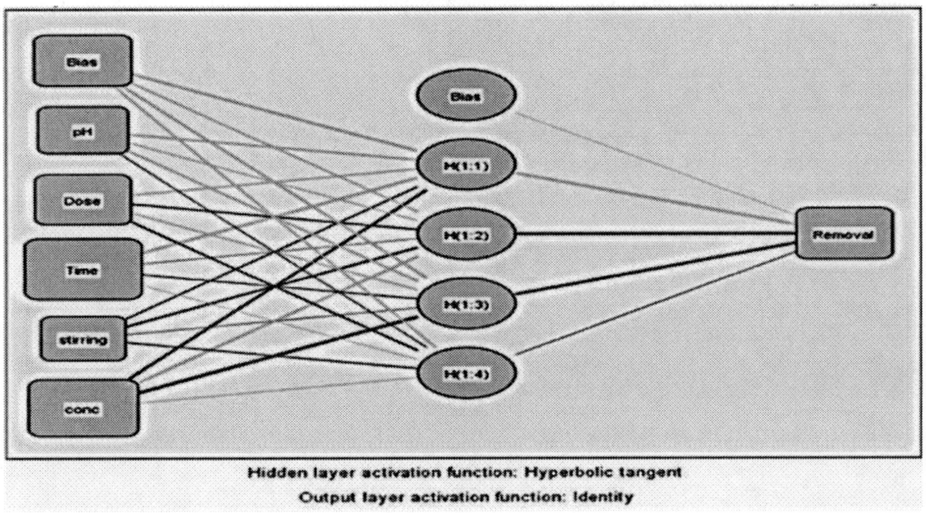

Figure 2. employing CA/nZVI, an ANN model for BOD removal is predicted.

3. Results and Discussion

3.1.Characterization of nZVI

The XRD result for powder nZVI with an angle (2) varying from 5 to 70 is shown in Figure 3. For the planes Fe (110) and Fe (200), respectively, the XRD pattern for the produced nZVI exhibits two major peaks at 2 equal 44.58° and 64.99°, respectively. In the produced powder samples, a distinct Fe peak indicated that nano-iron predominated. The Debye-Scherrer formula yielded a crystallite size range of 23 to 59 nm, which is consistent with the findings of the SEM examination, as shown in Figure 4a [81, 82]. Finally, Figure 4b showed that the sample with a tiny outer oxide surface was dominated by nano iron.

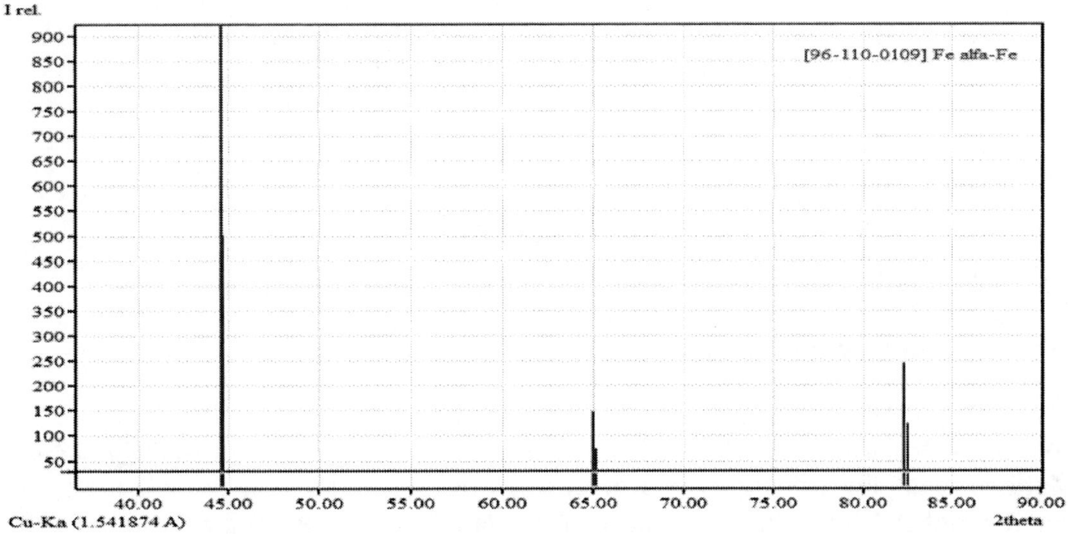

Figure 3. Results of X-ray powder diffraction for chemically produced nZVI

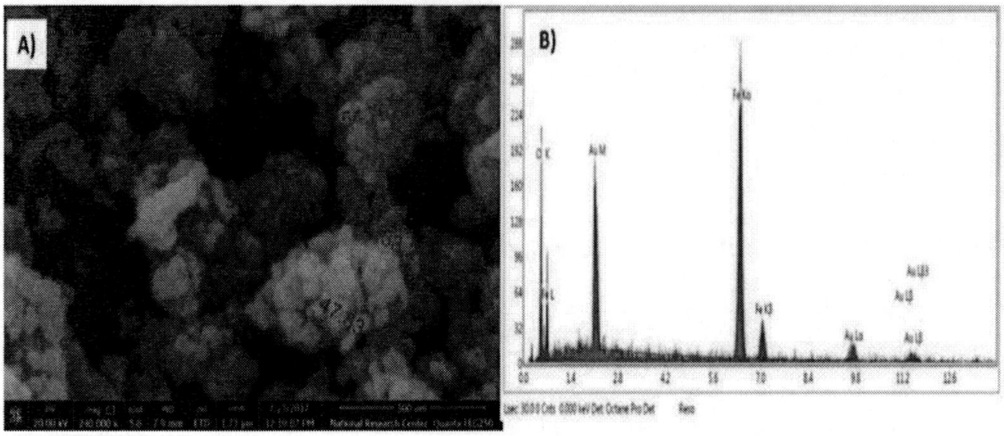

Figure 4. a) SEM examination of the produced nZVI particles; b) prepared nZVI findings from EDAX.

In order to explore the Point of Zero Charge (PZC), 0.1m of KCl was prepared, and then 20 mL were adjusted to pH values of 2, 3, 4, 5, 6, 7, 8, 9, 10, and 12 using 1 N H2SO4 or 1 N NaOH (pHi) and put into a 100 mL Erlenmeyer flask. The prepared flasks were filled with around 0.1 g of nZVI and kept at 23oC for 24 hours. The final pH readings were calculated (pHf). Five observations were used to calculate the average pH changes following nZVI, and all standard deviation values were within 0.1. By graphing the relationship between beginning pH values (pHi) and final pH values (final pH initial pHi), the PZC of nZVI was determined [49]. As shown in Fig. 5, the results showed that the PZC of nZVI was around 7.7, which is consistent with findings from earlier investigations.

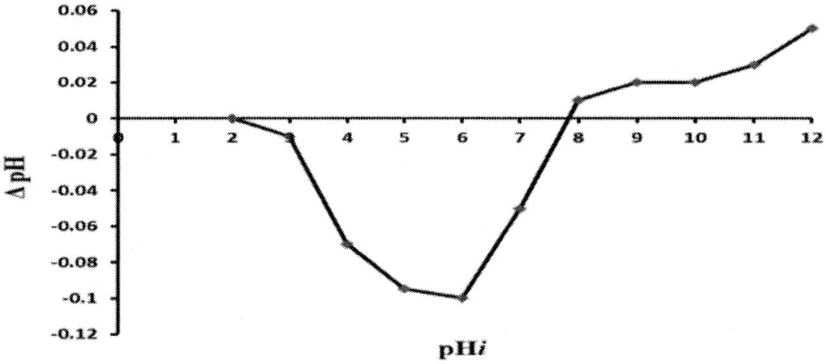

Figure 5. Point of Zero Charge of the nZVI freshly prepared powder.

3.2.Effect of operating conditions

3.2.1. Effect of pH

According to the batch study's explanation, the impact of pH was investigated for standard BOD solutions containing 300 mg/L in an acidic, neutral, and alkaline medium [81, 82]. According to Fig. 6a, the highest clearance % was seen at pH values between 6 and 8. Instead of the readily created free electrons that are responsible for the destruction of a wide range of biological contaminants, nanoparticles at the PZC may adsorb both positive and negative ions [32]. BOD removal from wastewater was investigated by Devi et al. in 2008 using an adsorption technique with an initial BOD concentration of 505 mg/L. They found that the greatest BOD concentration decrease occurred at an effective pH of 7.2, which is consistent with their findings [83]. Furthermore, according to many research, a pH between 6 and 9 is ideal for the highest elimination effectiveness [84]. BOD removal does not require pH adjustment, according to Mortula et al2012 .'s study "Removal of TDS and BOD from Synthetic Industrial Wastewater through Adsorption" [85]. Additionally, Zhang et al. (2010) investigated "Effect of limited artificial aeration on constructed wetland treatment of domestic wastewater," which looked at the significant impact of pH for the removal of 285mg/L BOD under various pH conditions. The results showed that there was no significant impact for any pH parameter with a p value greater than 0.05 [86].

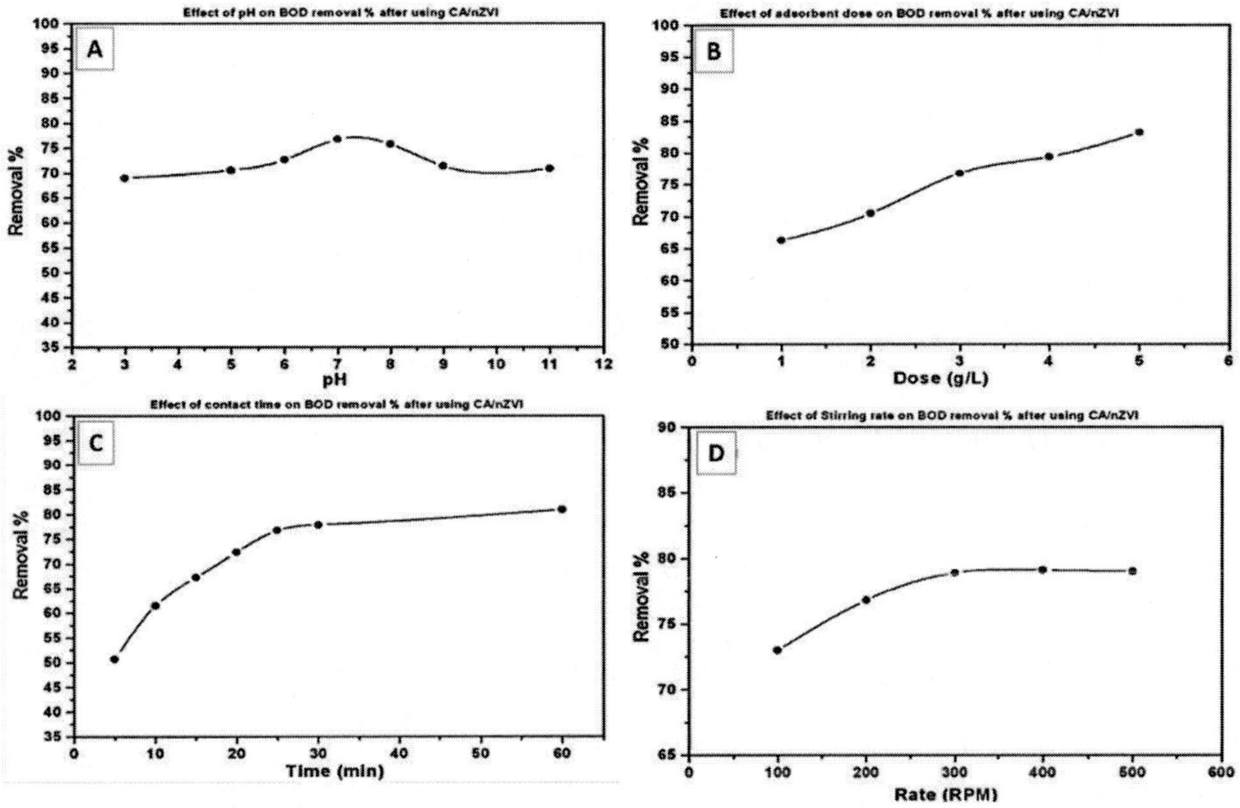

Figure 6. Effect of operating conditions for BOD removal: a) Effect of pH, B) Effect of Dose, Effect of contact time, and D) Effect of stirring rate.

3.2.2. Effect of Ag/nZVI dose

After utilizing CA/nZVI, the impact of adsorbent dosage on BOD removal is shown in Figure 6b. The outcomes showed that the removal efficiency was 76.8% and the lowest effective dosage of CA/nZVI adsorbent was 3g. Because there were more empty sites for adsorption and free electrons for the degradation process, the clearance efficiency rose with dosage. The removal percentage was 42% for baseline BOD values of 15001mg/L according to Laohaprapanon et al, 2010 .'s study on the removal of organic wastewater pollutants by wood fly ash utilizing a dosage of 160 g/L with a diameter less than 1mm [87]. For the initial BOD concentration of 12000 mg/L, Devi et al (2008).'s study on the reduction of BOD from coffee processing wastewater using avocado peel carbon revealed that the dose-effect is significantly significant for overall BOD reduction [88].

3.2.3. Effect of contact time

Figure 6c illustrates the influence of contact time on BOD elimination at various periods, including (5, 10, 15, 20, 25, 30, and 60min). At 25 minutes, the effective removal % was noted. For the starting BOD concentration of 1899.5 mg/L during 120 minutes, Akeem et al 2017's study on bio-derived MgO nano-powders for tannery BOD removal achieved a removal effectiveness of 96.9% [89]. BOD elimination occurs within 60 minutes, according to research done by Devi et al. in 2008 [88].

3.2.4. Effect of stirring rate

Figure 6d illustrates how stirring rate affects BOD elimination at various rates, including 100, 200, 300, 400, and 500 RPM. The outcomes showed that 200 RPM was the effective stirring rate. According to previous research, there is a little rise in removal percent in relation to the pace of chemisorption reaction stirring [90]. The large surface area and reactivity of the chosen nanomaterials are the primary drivers behind the use of nanoparticles for the degradation and adsorption of various biological pollutants at low stirring rates and times [91-95].

3.2.5. Effect of initial BOD concentrations

The impact of initial BOD concentration at various BOD concentrations, including 100, 200, 300, 400, and 500 mg/L, is depicted in Figure 7a. As shown in Fig. 7b, the obtained findings showed that the efficient BOD removal for the chosen AG/nZVI dosage was 300 mg/L with 76.8% removal. Devatha et al. (2016) used green synthesis nano iron recovered from leaf extracts to study home waste water. The removal effectiveness was 82% after employing FeNPs at pHs ranging from 6.5 to 8, which was in agreement with the results. [96].

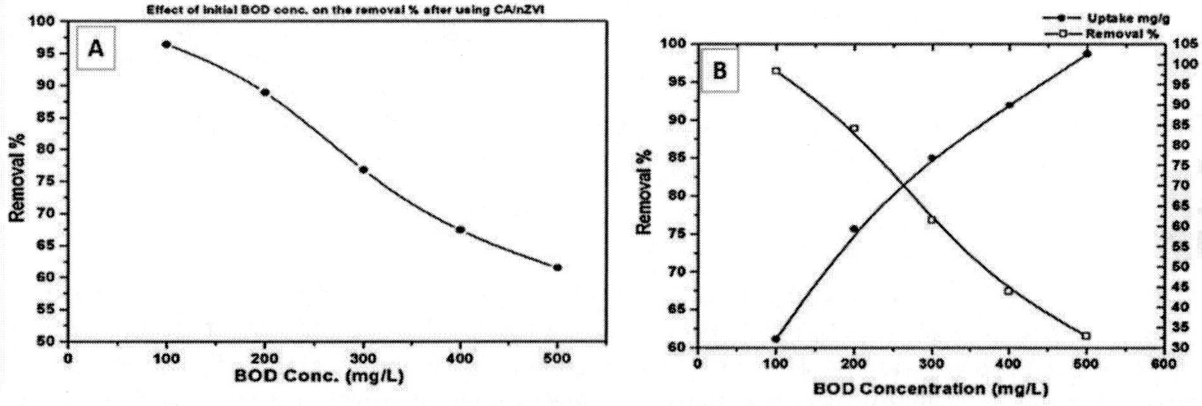

Figure 7. a) Effect of initial BOD concentrations, and B) Uptake relation

4. Isotherm studies

As illustrated in Figure 8, the nonlinear equations of the Toth, Redlich- Peterson, Khan, Hill, Sips, Koble-Corrigan, Javanovic, Langmuir, and Freundlich models were used to study the adsorption isotherms.

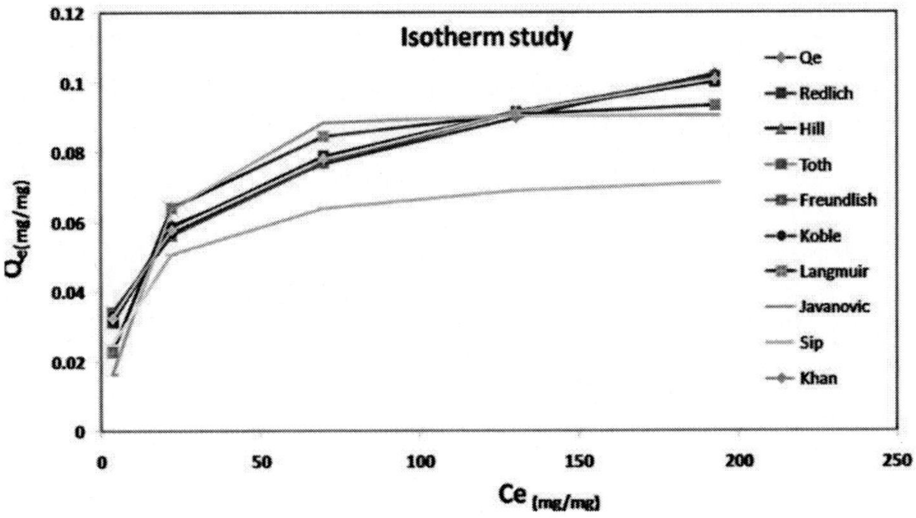

Figure 8. Adsorption isotherm for BOD removals using CA/nZVI.

The khan model is the most preferred model that can explain the adsorption process, according to the adsorption isotherm for the adsorption of BOD onto CA/nZVI. The maximal absorption of the adsorbent into solution, as determined by the Khan model, is 26 mg (BOD)/g (CA/nZVI), which explains the adsorption process for pure solutions.

Table 2. nonlinear adsorption isotherm for BOD removal using Ag/nZVI

	Redlich- Peterson		**Hill**		**Sips**		**Khan**		**Toth**	
Constants	Kr	0.032	QH	766.7	Qs	0.079	Qk	0.026	Kt	0.016
	Br	00.001	nH	0.274	Ks	0.100	Bk	1.211	at	12.212
	G	0.782	KD	31844	Bs	0.760	Ak	0.753	t	6.316
ERRORS										
Chi error	0.001		0.0013		0.0264		00.0011		0.0059	
ERRSQ	0.001		0.0010		0.0017		00.0010		0.0013	
HYBRD	0.001		0.0013		0.0196		00.0011		0.0048	
MPSD	0.002		0.0072		0.2490		0.0016		0.1117	
ARE	0.106		0.1434		1.0782		0.0852		0.5767	
EABS	0.008		0.0076		0.0810		0.0063		0.0321	
Error sum	0.116		0.1589		1.4559		**0.0934**		0.7314	

	Koble –Corrigan		**Javanovic**		**Freundlich**		**Langmuir**	
Constants	A	0.024	Q_m	0.090	Kf	0.024	Q0	0.099
	B	0.066	Kj	0.055	n	3.648	b	0.082
	D	0.338						
ERRORS								
Chi error	0.0012		0.0189		00.0013		0.0059	
ERRSQ	0.0010		0.0016		00.0010		00.0013	

HYBRD	0.0012	0.0114	00.0013	0.0048
MPSD	0.0036	0.2862	0.0072	0.1117
ARE	0.1234	0.8434	0.1435	0.5767
EABS	0.0079	0.0445	0.0076	0.0321
Error sum	0.1353	1.2048	0.1589	0.7314

5. Kinetic studies

The nonlinear relationships between several kinetic models are shown in Figure 9. The Pseudo First Order (P.F.O.), Pseudo Second Order (P.S.O.), Elovich, Avrami, and Intraparticle kinetic models were used to carry out the kinetic study.

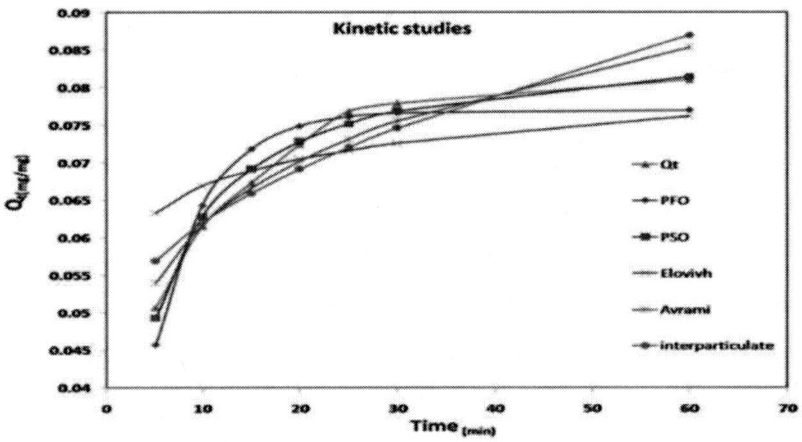

Figure 9. Kinetic studies for BOD removals using CA/nZVI.

The desorption of BOD onto CA/nZVI is reliant on concentration and dosage combined, and the adsorption mechanism is chemically rated controlling, according to the kinetic analysis for BOD removal using CA/nZVI accepted the P.S.O. kinetic mechanism. Indicated that the process is chemisorptions because the electrons are covalently shared or transferred between the sorbent and the sorbate. [90].

Table 3. nonlinear kinetic studies for BOD removal using CA/nZVI.

	P.F.O.	**P.S.O.**	**Elovich**	**Avrami**	**Intraparticle**
Constants	Qe = 0.077 K1 = 0.180	Qe = 0.087 K2 =3.061	α = 210.45 β = 193.109	Qe = 0.199 Kav = 0.219 Nav = 0.229	Kid=0.005 Ci=0.045
			ERRORS		
Chi error	0.001	0.001	0.004	0.001	0.002
ERRSQ	0.001	0.001	0.001	0.001	00.001
HYBRD	0.001	0.001	0.005	0.001	0.002
MPSD	0.020	0.003	0.083	0.011	0.029
ARE	0.320	0.121	0.583	0.241	0.376
EABS	0.021	0.008	0.037	0.017	0.025
Error sum	0.363	**0.132**	0.712	0.270	0.434

6. Response surface methodology (RSM)

The positive linear effect of the independent variables "dosage," "time," and "starting concentration" on BOD reduction was reported to be significant ($p < 0.05$) in the case of utilizing CA/nZVI, according to Table (4). However, the linear term of "pH" and "stirring rate" was shown to have no influence ($p > 0.05$). Table (4) contained both the corrected R^2 and the coefficient of determination (R^2) between measured data and simulated results. In order to define the most important operational parameters (p value equal to 0.001 contact duration and concentration) for BOD elimination utilizing CA/nZVI, the major result of the linear RSM model corresponds with the produced ANN model.

Equation 5 is applied to produce the removal equation. By decreasing or maximising the operational variables while employing all of the acquired results and not just those that are achieved under ideal circumstances, the removal equation may be utilized to manage the removal percentage.

$$Y = 65.158 + 1.364\, x_1 + 4.270\, x_2 + 0.497\, x_3 + 0.015\, x_4 - 0.091\, x_5 \qquad (6)$$

where, Y is the predicted response of different wastewater contaminants removal efficiency (%); $x_1$is pH (3 – 11); x_2 is adsorbent wet dose (1 – 5 g); x_3 is contact time (5 – 60 min); x_4 is stirring rate (100 – 500 rpm); x_5 is concentration (100 – 500 mg/L); β_0 is the model intercept; β_1, β_2, β_3, β_4 and β_5 are the linear coefficients of x_1, x_2, x_3, x_4 and x_5 respectively.

Table 4. *t*-statistics and p-values for coefficients of a linear regression model for BOD removal using CA/nZVI.

		RSM model (Enter Method)					
Model Summary	**R**	0.886					
	R^2	0.786					
	Adjusted R^2	0.726					
	Std. Error of the Estimate	4.948					
ANOVA	**F**	13.206					
	Model sig.	0.001					
Coefficients		**Term**	**Estimate**	**Standard Error**	**t-Ratio**	**Prob> |t|**	**Effect***
	Constant	β_0	65.158	10.555	6.173	0.001	Significant
	pH	β_1	1.364	1.145	1.191	.249	Insignificant
	Dose	β_2	4.270	1.565	2.729	.014	Significant
	Contact time	β_3	.497	.111	4.485	0.001	Significant
	Stirring rate	β_4	.015	.010	1.555	.137	Insignificant
	Concentration	β_5	-.091	.016	-5.83	0.001	Significant

7. ANNs

As indicated in Table (5) and Figure 2, the sample used in the CA/nZVI example was trained 14 times, tested 10 times, and ran a total of 24 times using the network structure 6-5-1.

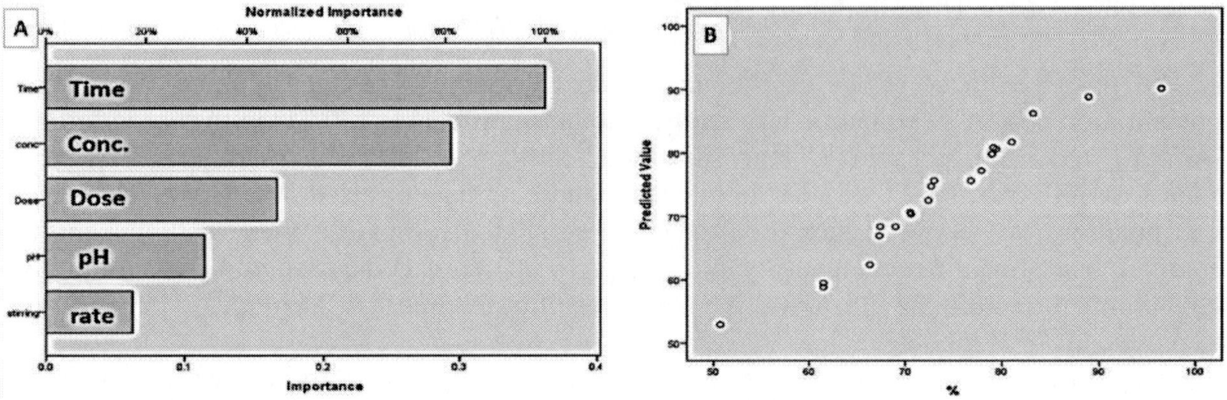

Figure 10. a) variable importance and b) relation between predicted and true values.

As shown in Figure 10a, the findings showed that Effect of contact time 100% and Effect of concentration 81.1% are the two most important operational parameters. As demonstrated in Figure 10b, the ANN findings showed a slight difference between the predicted and actual outcomes. Therefore, the residual percentage between the actual removal percentage and the projected removal percentage ranged between (-2.5, +7.5).

Table 5. Neural network results for BOD removal using CA/nZVI.

	Case processing summary		
	Parameter	BOD Removal ANN	
		N	**%**
	Sample training	14	58.3%
	Sample testing	10	41.7%
	Valid	24	100.0%
	Excluded	0	
	Total	**24**	
	Model summaries		
Training	Sum of Squares Error	.147	
	Relative Error	.023	
	Stopping Rule Used	1 consecutive step(s) with no decrease in error	
Testing	Sum of Squares Error	.476	
	Relative Error	.082	
	Independent Variable Importance		
		Importance	**Normalized importance**
	pH	.115	31.6%
	Dose	.167	46.0%
	Contact time	.362	100.0%
	Stirring rate	.063	17.3%
	Concentration	.294	81.1%

Conclusion

For initial BOD concentrations of 100, 200, 300, 400, and 500 mg/L, the impact of BOD removal by CA/nZVI was investigated, and the removal percentages were 96.4, 88.9, 76.8, 67.4, and 61.5%, respectively, after employing CA/nZVI. Wet CA/nZVI dosage (1, 2, 3, 4, and 5g), contact periods (5, 10, 15, 20, 25, 30, and 60min), stirring speeds (100, 200, 300, 400, and 500 rpm), and starting BOD concentrations (100, 200, 300, 400, and 500 mg/L) were all included for determining the effects of operational parameters. At pH 7, dosage 3 g/L, contact period 25 min, and stirring rate 200 rpm, the removal efficiency was at its highest. The nonlinear equations of the Redlich-Peterson, Hill, Sips, Khan, Toth, Koble-Corrigan, Javanovic, Freundlich, and Langmuir models were used to study the adsorption isotherms. According to the results, CA/nZVI preferred the Khan isotherm model, which had the lowest total of errors (0.0934). The P.S.O. kinetic model with the lowest error, 0.132, was favoured by CA/nZVI, according to the kinetic study. The most important factors, according to the ANN and RSM findings, were time, concentration, and dosage impacts.

References

1. Abdellatif G, Mahmoud AS, Peters RW, Mostafa MK: Waste Plastics and Microplastics in Africa: Negative Impacts and Opportunities. In *2021 AIChE Annual Meeting*AIChE; 2021
2. Mahmoud M, Mahmoud AS: Wastewater treatment using nano bimetallic iron/copper, adsorption isotherm, kinetic studies, and artificial intelligence neural networks. *Emergent Materials* 2021, 4:1455-1463.
3. Mahmoud M, Mahmoud AS, El-Said MA, Mostafa MK: Comparison of aluminum and iron nanoparticles for chromium removal from aqueous solutions and tannery wastewater, empirical modeling and prediction. *Emergent Materials* 2021:1-16.
4. Khopkar S: *Environmental pollution monitoring and control.* New Age International; 2007.
5. Mahmoud M, Mahmoud ASJEM: Wastewater treatment using nano bimetallic iron/copper, adsorption isotherm, kinetic studies, and artificial intelligence neural networks. 2021, 4:1455-1463.
6. Wang J-H, Mahmoud MS, Mahmoud ASJB: Integrated Efficiency of Using Nanocellulose-Nano Zero Valent Iron Composite in Water Treatment. 2022, 17:975-992.
7. Hamdy A, Mostafa MK, Nasr MJES, Research P: Techno-economic estimation of electroplating wastewater treatment using zero-valent iron nanoparticles: batch optimization, continuous feed, and scaling up studies. 2019, 26:25372-25385.
8. Kurniawan TA, Sillanpää ME, Sillanpää M: Nanoadsorbents for remediation of aquatic environment: local and practical solutions for global water pollution problems. *Critical reviews in environmental science and technology* 2012, 42:1233-1295.
9. Maha M. Elshfai ASM, Elsaid MA: Comparative Studies of using nZVI and Entrapped nZVI in Alginate Biopolymer (Ag/ nZVI) for Aqueous Phosphate Removal. In *12th international conference on nano-technology for green and sustainable construction*https://www.researchgate.net/publication …; 2021: 21.
10. Mahmoud M, Mahmoud AS, El-Said MA, Mostafa MKJEM: Comparison of aluminum and iron nanoparticles for chromium removal from aqueous solutions and tannery wastewater, empirical modeling and prediction. 2021:1-16.

11. Edberg S, Rice E, Karlin R, Allen M: Escherichia coli: the best biological drinking water indicator for public health protection. *Journal of applied microbiology* 2000, 88:106S-116S.
12. Mata TM, Martins AA, Caetano NS: Microalgae for biodiesel production and other applications: a review. *Renewable and sustainable energy reviews* 2010, 14:217-232.
13. Wright R: The survival patterns of selected faecal bacteria in tropical fresh waters. *Epidemiology & Infection* 1989, 103:603-611.
14. Anderson KL, Whitlock JE, Harwood VJ: Persistence and differential survival of fecal indicator bacteria in subtropical waters and sediments. *Appl Environ Microbiol* 2005, 71:3041-3048.
15. Tyagi V, Chopra A, Kazmi A, Kumar A: Alternative microbial indicators of faecal pollution: current perspective. *Journal of Environmental Health Science & Engineering* 2006, 3:205-216.
16. Stentiford G, Longshaw M, Lyons B, Jones G, Green M, Feist S: Histopathological biomarkers in estuarine fish species for the assessment of biological effects of contaminants. *Marine Environmental Research* 2003, 55:137-159.
17. Carballa M, Omil F, Lema JM, Llompart Ma, García-Jares C, Rodríguez I, Gomez M, Ternes T: Behavior of pharmaceuticals, cosmetics and hormones in a sewage treatment plant. *Water research* 2004, 38:2918-2926.
18. Akpor OB, Ohiobor GO, Olaolu TD: Heavy metal pollutants in wastewater effluents: sources, effects and remediation. *Advances in Bioscience and Bioengineering* 2014, 2:37-43.
19. DeRegnier DP, Cole L, Schupp DG, Erlandsen SL: Viability of Giardia cysts suspended in lake, river, and tap water. *Appl Environ Microbiol* 1989, 55:1223-1229.
20. Ortega YR, Adam RD: Giardia: overview and update. *Clinical infectious diseases* 1997, 25:545-549.
21. Adam RD: Biology of Giardia lamblia. *Clinical microbiology reviews* 2001, 14:447-475.
22. Mahmoud AS, SaryEl-deen RA, Mostafa MK, Peters RW: Artificial Intelligence for Organochlorine Pesticides Removal from Aqueous Solutions using Entrapped nZVI in Alginate Biopolymer. In *2017 Annual AIChE Meeting; Minneapolis, MN, October 29 - November 3, 2017*2017
23. Pronk M, Goldscheider N, Zopfi J: Particle-size distribution as indicator for fecal bacteria contamination of drinking water from karst springs. *Environmental Science & Technology* 2007, 41:8400-8405.
24. White WB: Hydrogeology of karst aquifers. In *Encyclopedia of caves.* Elsevier; 2019: 537-545
25. Shon H, Vigneswaran S, Kandasamy J, Cho J: Characteristics of effluent organic matter in wastewater. *Eolss, Oxford* 2007.
26. Ellis T: Chemistry of wastewater. Encyclopedia of Life Support System (EOLSS), Developed under the Auspices of the UNESCO. Eolss Publishers, Oxford, UK, http://www. eolss. net; 2004.
27. Verlicchi P, Al Aukidy M, Zambello E: Occurrence of pharmaceutical compounds in urban wastewater: removal, mass load and environmental risk after a secondary treatment—a review. *Science of the total environment* 2012, 429:123-155.
28. M Abdeldayem O, Eldaghar O, K Mostafa M, M Habashy M, Hassan AA, Mahmoud H, M Morsy K, Abdelrady A, Peters RWJW: Mitigation plan and water harvesting of flashflood in arid rural communities using modelling approach: A case study in Afouna village, Egypt. 2020, 12:2565.

29. Morsy KM, Mostafa MK, Abdalla KZ, Galal MMJA, Soil, Research W: Life cycle assessment of upgrading primary wastewater treatment plants to secondary treatment including a circular economy approach. 2020, 13:1178622120935857.
30. Su Z-H, Mahmoud MS, Fan S, Zhang Y, Peng FJB: Combustion properties of mixed black liquor solids from linter and reed pulping. 2019, 14:8278-8288.
31. Anjum M, Miandad R, Waqas M, Gehany F, Barakat M: Remediation of wastewater using various nano-materials. *Arabian Journal of Chemistry* 2019, 12:4897-4919.
32. Mahmoud ASJhwrnpAMpStromhpfasunledeS: Study the Removal of Most Hazardous Pollutants from Aqueous Solutions Using Nanoparticles. http://lis.cl.cu.edu.eg/search*ara; 2017.
33. Rabie S. Farag MME-S, Ahmed S. Mahmoud, Mohamed K. Mostafa,, Peters RW: Green Synthesis of Nano Iron Carbide: Preparation, Characterization and Application for Removal of Phosphate from Aqueous Solutions. In *Proc 2018 Annual AIChE Meeting, Pittsburgh, PA, (October 28 – November 2)*2018
34. Mahmoud AS, Farag RS, Elshfai MM, Mohamed LA, Ragheb SMJA, Soil, Research W: Nano zero-valent aluminum (nZVAl) preparation, characterization, and application for the removal of soluble organic matter with artificial intelligence, isotherm study, and kinetic analysis. 2019, 12:1178622119878707.
35. Saryel-Deen RA, Mahmoud AS, Mahmoud M, Mostafa MK, Peters RW: Adsorption and kinetic studies of using entrapped sewage sludge ash in the removal of chemical oxygen demand from domestic wastewater, with artificial intelligence approach. In *2017 Annual AIChE Meeting*2017
36. Fan S, Mahmoud MS, Wen B, Su Z, Zhang YJB: Bioelectric activity of microbial fuel cell during treatment of old corrugated containerboard discharges. 2018, 13:3545-3553.
37. Ferroudj N, Nzimoto J, Davidson A, Talbot D, Briot E, Dupuis V, Bée A, Medjram MS, Abramson S: Maghemite nanoparticles and maghemite/silica nanocomposite microspheres as magnetic Fenton catalysts for the removal of water pollutants. *Applied Catalysis B: Environmental* 2013, 136:9-18.
38. El-Shafei M, Mahmoud A, Mostafa M, Peters R: Effects of entrapped nZVI in alginate polymer on BTEX removal. In *AIChE Annual Meeting San Francisco, CA*2016: 13-18.
39. Mostafa MK, Mahmoud AS, SaryEl-deen RA, Peters RW: Application of Entrapped Nano Zero Valent Iron into Cellulose Acetate Membranes for Domestic Wastewater Treatment. In *2017 Annual AIChE Meeting; Minneapolis, MN, October 29 - November 3, 2017*2017
40. Tyagi I, Gupta V, Sadegh H, Ghoshekandi RS, Makhlouf AH: Nanoparticles as adsorbent; a positive approach for removal of noxious metal ions: a review. *Science Technology and Development* 2017, 34:195-214.
41. Farag RS, Elshfai MM, Mahmoud ASJJoE, Research B: Adsorption and kinetic studies using nano zero valent iron (nZVI) in the removal of chemical oxygen demand from aqueous solution with response surface methodology and artificial neural network approach. 2018, 7:12-22.
42. Mahmoud AS, EL-Tayieb MM, Ahmed NAS, Mostafa AMJJoE, Research B: Algorithms and statistics for municipal wastewater treatment using nano zero valent iron (nZVI). 2018, 7:30-44.
43. Kyzas GZ, Matis KA: Nanoadsorbents for pollutants removal: a review. *Journal of Molecular Liquids* 2015, 203:159-168.

44. Heberling JA, Adewuyi YG, Mahmoud AS, Mostafa MK, Peters R: AOP Performance at Wastewater Treatment Plants. In *2018 AIChE Annual Meeting*AIChE; 2018
45. Varnosfaderany MN, Mirghaffary N, Ebrahimi E, Soffianian A: Water quality assessment in an arid region using a water quality index. *Water Science and Technology* 2009, 60:2319-2327.
46. Jin P, Jin X, Wang XC, Shi X: An analysis of the chemical safety of secondary effluent for reuse purposes and the requirement for advanced treatment. *Chemosphere* 2013, 91:558-562.
47. Ahmed Karam MKM, Abdelsalam Elawwad, Khaled Zaher, Ahmed S. Mahmoud, Peters RW: Semi-Pilot Plant for Tertiary Treatment of Domestic Wastewater using Algal Photo-Bioreactor, with Artificial Intelligence. In *2019 Annual AIChE Meeting; Orlando, FL, November 10 - 15, 2019*https://www.researchgate.net/publication/337403206_562bb_Semi …; 2019
48. Mahmoud AS, Mostafa MK, Peters RW: A prototype of textile wastewater treatment using coagulation and adsorption by Fe/Cu nanoparticles: Techno-economic and scaling-up studies. *Nanomaterials and Nanotechnology* 2021, 11:18479804211041181.
49. Rabie S. Farag MME, Ahmed S. Mahmoud, Mohamed K. Mostafa, Ahmed Karam,, Peters RW: Study the Degradation and Adsorption Processes of Organic Matters from Domestic Wastewater using Chemically Prepared and Green Synthesized Nano Zero-Valent Iron. In *2019 Annual AIChE Meeting; Orlando, FL, November 10 - 15, 2019*https://www.researchgate.net/publication …; 2019
50. Thakur VK, Voicu SI: Recent advances in cellulose and chitosan based membranes for water purification: a concise review. *Carbohydrate polymers* 2016, 146:148-165.
51. Miculescu M, Thakur VK, Miculescu F, Voicu SI: Graphene-based polymer nanocomposite membranes: a review. *Polymers for Advanced Technologies* 2016, 27:844-859.
52. Voicu S, Ninciuleanu C, Muhulet O, Miculescu M: Cellulose acetate membranes with controlled porosity and their use for the separation of aminoacids and proteins. *J Optoelectron Adv Mater* 2014, 16:903-908.
53. Thakur VK, Vennerberg D, Madbouly SA, Kessler MR: Bio-inspired green surface functionalization of PMMA for multifunctional capacitors. *RSC Advances* 2014, 4:6677-6684.
54. Thakur V, Singha A, Thakur M: Surface modification of natural polymers to impart low water absorbency. *International Journal of Polymer Analysis and Characterization* 2012, 17:133-143.
55. Azzaoui K, Mejdoubi E, Lamhamdi A, Zaoui S, Berrabah M, Elidrissi A, Hammouti B, Fouda MM, Al-Deyab SS: Structure and properties of hydroxyapatite/hydroxyethyl cellulose acetate composite films. *Carbohydrate polymers* 2015, 115:170-176.
56. Azzaoui K, Lamhamdi A, Mejdoubi EM, Berrabah M, Hammouti B, Elidrissi A, Fouda MM, Al-Deyab SS: Synthesis and characterization of composite based on cellulose acetate and hydroxyapatite application to the absorption of harmful substances. *Carbohydrate polymers* 2014, 111:41-46.
57. Hokkanen S, Bhatnagar A, Repo E, Lou S, Sillanpää M: Calcium hydroxyapatite microfibrillated cellulose composite as a potential adsorbent for the removal of Cr (VI) from aqueous solution. *Chemical Engineering Journal* 2016, 283:445-452.
58. Hokkanen S, Repo E, Westholm LJ, Lou S, Sainio T, Sillanpää M: Adsorption of Ni2+, Cd2+, PO43− and NO3− from aqueous solutions by nanostructured microfibrillated cellulose modified with carbonated hydroxyapatite. *Chemical Engineering Journal* 2014, 252:64-74.

59. Zhong Z, Qin J, Ma J: Cellulose acetate/hydroxyapatite/chitosan coatings for improved corrosion resistance and bioactivity. *Materials Science and Engineering: C* 2015, 49:251-255.
60. Eftekhari S, El Sawi I, Bagheri ZS, Turcotte G, Bougherara H: Fabrication and characterization of novel biomimetic PLLA/cellulose/hydroxyapatite nanocomposite for bone repair applications. *Materials Science and Engineering: C* 2014, 39:120-125.
61. Salama A, El-Sakhawy M: Regenerated cellulose/wool blend enhanced biomimetic hydroxyapatite mineralization. *International journal of biological macromolecules* 2016, 92:920-925.
62. Lan T, Shao Z-q, Wang J-q, Gu M-j: Fabrication of hydroxyapatite nanoparticles decorated cellulose triacetate nanofibers for protein adsorption by coaxial electrospinning. *Chemical Engineering Journal* 2015, 260:818-825.
63. Garai S, Sinha A: Biomimetic nanocomposites of carboxymethyl cellulose–hydroxyapatite: novel three dimensional load bearing bone grafts. *Colloids and Surfaces B: Biointerfaces* 2014, 115:182-190.
64. Mahmoud AS, Farag RS, Elshfai MM: Reduction of organic matter from municipal wastewater at low cost using green synthesis nano iron extracted from black tea: Artificial intelligence with regression analysis. *Egyptian Journal of Petroleum* 2020, 29:9-20.
65. Mahmoud M, Mahmoud AS: Wastewater treatment using nano bimetallic iron/copper, adsorption isotherm, kinetic studies, and artificial intelligence neural networks. *Emergent Materials* 2021:1-9.
66. Mahmoud M: Decolorization of certain reactive dye from aqueous solution using Baker's Yeast (Saccharomyces cerevisiae) strain. *HBRC journal* 2016, 12:88-98.
67. Baird RB: Standard methods for the examination of water and wastewater, 23rd. Water Environment Federation, American Public Health Association, American …; 2017.
68. Mahmoud AS, Mostafa MK, Abdel-Gawad SA: Artificial intelligence for the removal of benzene, toluene, ethyl benzene and xylene (BTEX) from aqueous solutions using iron nanoparticles. *Water Supply* 2018, 18:1650-1663.
69. Mahmoud AS, EL-Tayieb MM, Ahmed NAS, Mostafa AM: Algorithms and statistics for municipal wastewater treatment using nano zero valent iron (nZVI). *Journal of Environment and Biotechnology Research* 2018, 7:30-44.
70. Abdel-Gawad SA, Baraka AM, El-Shafei MM, Mahmoud AS: Effects of nano zero valent iron and entrapped nano zero valent iron in alginate polymer on poly aromatic hydrocarbons removal. *Journal of Environment & Biotechnology Research* 2016, 5:18-28.
71. Mahmoud AS, Mohamed NY, Mostafa MK, Mahmoud MS: Effective chromium adsorption from aqueous solutions and tannery wastewater using bimetallic Fe/Cu nanoparticles: response surface methodology and artificial neural network. *Air, Soil and Water Research* 2021, 14:11786221211028162.
72. Mahmoud AS: Effect of nano bentonite on direct yellow 50 dye removal; Adsorption isotherm, kinetic analysis, and thermodynamic behavior. *Progress in Reaction Kinetics and Mechanism* 2022, 47:14686783221090377.
73. Mahmoud MS, Mahmoud AS: Treatment of Chromium from Tannery Wastewater Using Nano-Bio Polymer Complex. In *Key Engineering Materials*Trans Tech Publ; 2022: 191-203.
74. Abdel-Gawad SA, Baraka AM, El-Shafei MM, Mahmoud ASJJoE, Research B: Effects of nano zero valent iron and entrapped nano zero valent iron in alginate polymer on poly aromatic hydrocarbons removal. 2016, 5:18-28.

75. Mahmoud AS, Ismail A, Mostafa MK, Mahmoud M, Ali W, Shawky AM: Isotherm and kinetic studies for heptachlor removal from aqueous solution using Fe/Cu nanoparticles, artificial intelligence, and regression analysis. *Separation Science and Technology* 2020, 55:684-696.
76. Kamar MT, Elattar H, Mahmoud AS, Peters RW, Mostafa MK: A critical review of state-of-the-art technologies for electroplating wastewater treatment. *International Journal of Environmental Analytical Chemistry* 2022:1-34.
77. Mostafa MK, Mahmoud AS, Mahmoud MS, Nasr M: Computational-Based Approaches for Predicting Biochemical Oxygen Demand (BOD) Removal in Adsorption Process. *Adsorption Science & Technology* 2022, 2022.
78. Mahmoud AS, Ismail A, Mostafa MK, Mahmoud M, Ali W, Shawky AMJSS, Technology: Isotherm and kinetic studies for heptachlor removal from aqueous solution using Fe/Cu nanoparticles, artificial intelligence, and regression analysis. 2020, 55:684-696.
79. Mahmoud AS, Mostafa MK, Abdel-Gawad SAJWS: Artificial intelligence for the removal of benzene, toluene, ethyl benzene and xylene (BTEX) from aqueous solutions using iron nanoparticles. 2018, 18:1650-1663.
80. Mahmoud AS, Mostafa MK, Nasr MJSS, Technology: Regression model, artificial intelligence, and cost estimation for phosphate adsorption using encapsulated nanoscale zero-valent iron. 2019, 54:13-26.
81. Mahmoud AS, Mahmoud MS, Yousri N, Mostafa MK: Advanced Wastewater Treatment by Using Entrapped nZVI into Alginate (Ag) Biopolymer: Adsorption Isotherm, Kinetic Models, and Statistical Analysis. In *Key Engineering Materials*Trans Tech Publ; 2022: 147-160.
82. Sary El-Deen RA, Mahmoud NY, Mahmoud AS: Thermodynamic Equilibrium Kinetics of Sewage Sludge Ash Pellets for the Removal of Methylene Blue (MB) from Aqueous Solutions. In *Key Engineering Materials*Trans Tech Publ; 2022: 161-172.
83. Devi R, Dahiya R: COD and BOD removal from domestic wastewater generated in decentralised sectors. *Bioresource Technology* 2008, 99:344-349.
84. Jung J-Y, Lee S-M, Shin P-K, Chung Y-C: Effect of pH on phase separated anaerobic digestion. *Biotechnology and Bioprocess Engineering* 2000, 5:456.
85. Mortula M, Shabani S: Removal of TDS and BOD from synthetic industrial wastewater via adsorption. In *International Conference of Chemical, Biological & Environmental Engineering, Singapore*2012
86. Zhang L-y, Zhang L, Liu Y-d, Shen Y-w, Liu H, Xiong Y: Effect of limited artificial aeration on constructed wetland treatment of domestic wastewater. *Desalination* 2010, 250:915-920.
87. Laohaprapanon S, Marques M, Hogland W: Removal of Organic Pollutants from Wastewater Using Wood Fly Ash as a Low-Cost Sorbent. *CLEAN–Soil, Air, Water* 2010, 38:1055-1061.
88. Devi R, Singh V, Kumar A: COD and BOD reduction from coffee processing wastewater using Avacado peel carbon. *Bioresource technology* 2008, 99:1853-1860.
89. Oladipo AA, Adeleye OJ, Oladipo AS, Aleshinloye AO: Bio-derived MgO nanopowders for BOD and COD reduction from tannery wastewater. *Journal of water process engineering* 2017, 16:142-148.
90. Ho Y-S, McKay G: Pseudo-second order model for sorption processes. *Process biochemistry* 1999, 34:451-465.
91. Machado S, Stawiński W, Slonina P, Pinto A, Grosso J, Nouws H, Albergaria JT, Delerue-Matos C: Application of green zero-valent iron nanoparticles to the remediation of soils contaminated with ibuprofen. *Science of the Total Environment* 2013, 461:323-329.

92. Mahdy SA, Raheed QJ, Kalaichelvan P: Antimicrobial activity of zero-valent iron nanoparticles. *International Journal of Modern Engineering Research* 2012, 2:578-581.
93. Markova Z, Šišková KnM, Filip J, Čuda J, Kolář M, Šafářová Kr, Medřík I, Zbořil R: Air stable magnetic bimetallic Fe–Ag nanoparticles for advanced antimicrobial treatment and phosphorus removal. *Environmental science & technology* 2013, 47:5285-5293.
94. Yang J, Wang X, Zhu M, Liu H, Ma J: Investigation of PAA/PVDF–NZVI hybrids for metronidazole removal: synthesis, characterization, and reactivity characteristics. *Journal of hazardous materials* 2014, 264:269-277.
95. Yang Y, Guo J, Hu Z: Impact of nano zero valent iron (NZVI) on methanogenic activity and population dynamics in anaerobic digestion. *Water research* 2013, 47:6790-6800.
96. Devatha C, Thalla AK, Katte SY: Green synthesis of iron nanoparticles using different leaf extracts for treatment of domestic waste water. *Journal of cleaner production* 2016, 139:1425-1435.

533n Remote Monitoring, Supervisory Control and Technoeconomic Evaluation of Advanced High Recovery Wellhead Water Purification and Desalination Systems

Yoram Cohen, Bilal Khan, Christian Aguilar, Mitchell Rogers, Yang Zhou, Nora Marki

Remote monitoring, supervisory control and technoeconomic evaluation of advanced high recovery wellhead water purification and desalination systems

Increased nitrate contamination and salinity of potable well water supplies in agricultural regions throughout the United States is adversely impacting many small and disadvantaged communities. While wellhead membrane-based nitrate removal and salinity reduction is feasible, small communities lack the technical resources to operate and manage water treatment, particularly when confronted with temporal variability of water quality and community water use patterns. Thus, the deployment of water treatment systems requires advanced self-adaptive operation and remote monitoring and supervision. The goal is to reduce the cost of onsite operational and monitoring needs, relying on expert operators that are remotely located with the tools afforded via real-time cloud-computing, online and remote system monitoring, and advanced supervisory control that allows adherence to regulatory water quality and other small water system requirements. In the current work, an advanced design of a high recovery membrane-based water treatment and desalination system is introduced, focusing on local and remote monitoring requirements, self-adaptive operation and technoeconomic feasibility analysis. The study is based on operational pilot study experience with our uniquely designed high recovery reverse osmosis (RO) systems, integrated with feedwater pretreatment and product water post-processing, deployed to provide safe drinking water to a number of small, disadvantaged communities in an agricultural region in California. Through remote system monitoring and supervisory control, the concept of a "virtual water district" is introduced, whereby multiple small community water systems can be managed and operated remotely. Here we demonstrate the utility of modern advances in wireless communication, cloud-based database management, and "live" user/operator interfaces. Managing of the treatment system is facilitated via a cyberinfrastructure consisting of three major components: local system control, automated cloud-based data collection, and a web-based interface that provide real-time monitoring and historical data from each of the water systems, in addition to providing alerts and streaming regulatory data for these systems that currently provide safe drinking water to three small communities. Intermittent operation of the water systems is facilitated via a supervisory control system to enable meeting the temporally varying water use of the communities while ensuring compliance with regulatory requirements regarding water quality. Examples of system operation, and approaches to mitigate membrane fouling and mineral scaling will be discussed within the context of expanding the concept of remotely managed distributed water systems. Ina addition, techno-economic assessment was conducted to determine the water production cost for these small wellhead water treatment systems based on actual expenditures and process data over a two year operating period. Results of the present pilot study demonstrated that advanced remotely operated, distributed membrane-based water treatment systems are an economically viable solution for providing impacted disadvantaged communities with safe drinking water.

533q Cationic Dye Adsorption on Metal Organic Framework: An Equilibrium Study

Saif Mehdi, Kannan Aravamudan

Environmental pollution is a huge concern which is complicated by different types of pollutants present in aqueous streams. The pollutants include toxic organics, dyes, heavy metal ions, pharmaceutical compounds, etc. In this study, we focus on dyes present in textile industry effluents. These dyes, even at low concentrations cause coloring of water streams leading them to lose their aesthetic appeal. Further, sunlight penetration in water bodies is hindered leading to disruption in the marine ecosystem. These dyes are difficult to be broken down by biological processes and, hence, suitable alternatives are required. Adsorption is the preferred choice owing to its economy, simplicity in design and operation and ability to be effective even at very low concentrations. Adsorbent materials are a hotbed of immense research activity where there is considerable scope for synthesizing materials with pollutant capacity enhancement and rapid uptake. A new class of adsorbents termed as Metal Organic Frameworks (MOF) is gaining popularity in recent times. These classes of adsorbents have been found to be having high BET surface area, tuneable porosity and functionality which enable in deep removal of toxic pollutants in aqueous streams.

In the present work, a functionalized MOF (MIL-101-SO_3H[1]) was synthesized hydrothermally and characterized for its physico-chemical properties such as BET surface area, morphology, surface composition and point of zero-charge. MIL-SO_3H (MILS) was used to adsorb a cationic dye viz. basic yellow. Optimal conditions for enhanced equilibrium capacities were identified using Design of Experiments (DoE). The adsorption capacity was found to be 320 mg/g at pH 6, temperature 25 °C, and dosage of 1 g/L. The equilibrium data was fitted to Simple Hill's Model (SHM), a statistical physics-based model which gives considerable insight into the interaction between adsorptive and adsorbent. The adsorption process was found to be exothermic with enthalpy of adsorption of -29.37 kJ/mol. The adsorption mechanism was revealed to be an electrostatic interaction between the cationic dye and the SO_3^- group present on the functionalized MOF. Kinetic studies were also carried out with this MOF at the optimal conditions.

References:

1. Ma, L., Xu, L., Jiang, H. andamp; Yuan, X. Comparative research on three types of MIL-101(Cr)-SO 3 H for esterification of cyclohexene with formic acid. *RSC Adv.* **9**, 5692–5700 (2019).

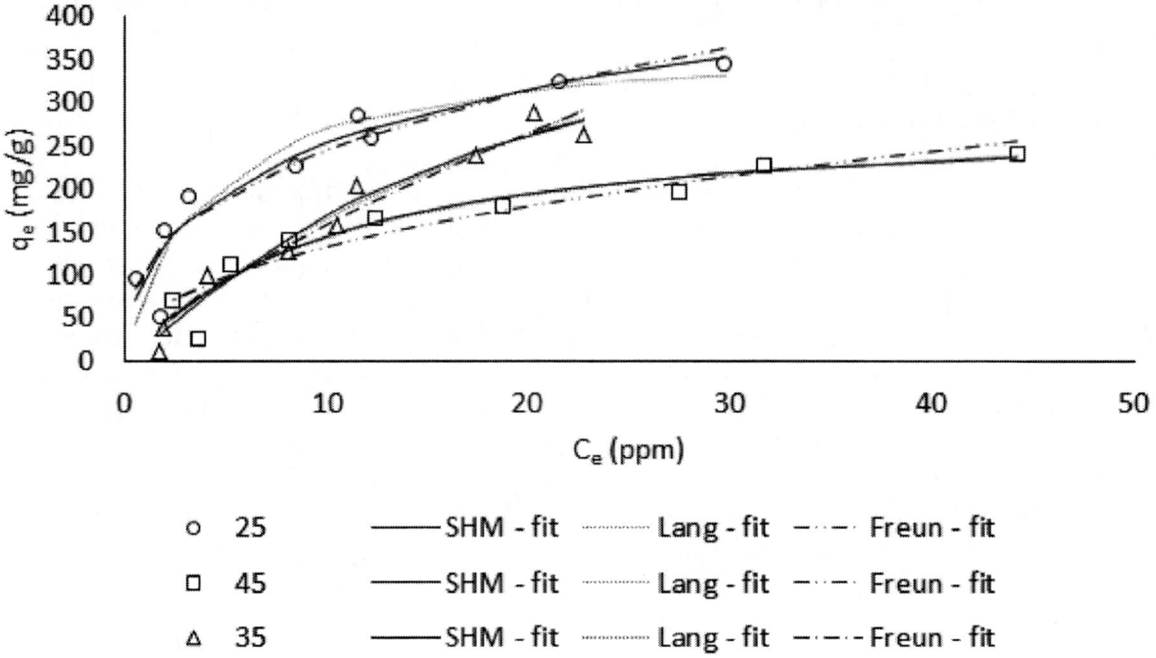

400
350
300
250
200
150
100
50
0
q_e (mg/g)
0
10
20
30
40
50
C_e (ppm)
25
SHM - fit
Lang - fit
Freun - fit
45
SHM - fit
Lang - fit
Freun - fit
35
SHM - fit
Lang - fit
Freun - fit

533r A Comprehensive Study to Evaluate the Removal Capacity of Heavy Metal Ions and Dissolved Organics from Produced Water Using Vadose Zone Soils for Soil Aquifer Treatment

Jisha Ali, Tu Pham Le Phuong, Abdulfahim Arangadi, Emad Alhseinat

Produced water is one of the largest streams of wastewater generated from the oil and gas industries and its production is exponentially increasing in the upcoming years due to the unpredictable increase in the demand for oil and gas over the world. Since produced water is acting as complex contaminated toxic slurry because of the presence of dissolved and dispersed oil components, production chemicals, dissolved formation minerals, produced solids, and dissolved gases (including CO_2 and H_2S), it displayed detrimental effects on the surface, soil, aquatic system, and underground water as this often discharged to surface water without appropriate treatment. However, the nature of contaminants and their percentages constituting the produced water can vary drastically based on the geolocation and the type of petroleum products. The treatment of produced water is now very important due to the treatment will not only decline the drastic environmental effects of the industry, moreover, it takes the edge off the stress on freshwater resources. As a result of the great demand for oil and gas, a large quantity of produced water is generated, hence major problem facing the treatment of produced water should need to handle large volumes with covering the removal of a range of contaminants. The UAE is one of the countries facing the problem of high production of produced water and its treatment as it is very rich in oil and gas industries. Currently, soil aquifer treatment (SAT) is an attractive method in the area of wastewater treatment in developed and developing countries because of its advantages such as removing multiple contaminants, being robust, environment-friendly, and minimum use of energy and chemicals. SAT denotes the artificial recharge or infiltration of wastewater through the unsaturated zone (vadose zone) to boost the underlying aquifers. Since, SAT improves the physical, chemical, and microbial quality of contaminated water by removing particles during soil passage, as well as provides long-term storage without evaporative losses and less external re-contamination, it can thus substitute or support other water treatment methods. But the major limitation of this method is the leaching of aquifer materials such as iron, manganese, calcium, selenium, arsenic, chromium or fluoride, etc in extracted water under reducing conditions. Thus, a detailed study on the behaviour or interaction of soils from the vadose zone with contaminated water, and the investigation on the mechanism of removal should be conducted for the SAT treatment method before it becomes a part of the water treatment process or independent treatment. Since, SAT can handle the treatment of large volume of contaminated water as well as environmentally friendly and cost effective method, it can be applied for the treatment of produced water from oil and gas industries. The present study focussed on the suitability of four soils such as sand, sandstone, siltstone, and conglomerate which were collected from the vadose zone of Abu Dhabi, UAE, for the treatment of produced water for SAT method. This comprehensive study was aimed to monitor the removal capacity on two heavy metals (Ni (II), and Zn (II)), and organic contaminants such as phenol with the same concentration as their in produced water using selected four soil samples. The four soil samples were utilized without further treatment or modification in order to keep a cost-effective approach. The detailed characterizations of soil samples were conducted using different techniques (SEM, EDX, XRF, FTIR, and XRD analysis) to analyze their chemical composition and physicochemical properties. The results found that sandstone and conglomerate were rich in calcium carbonate-based minerals, and sand and siltstone were based on silica-based minerals along with other minerals. To understand the interaction between metal ions/phenol and soil samples, batch experiments were conducted on a laboratory scale at normal pH at room temperature for two weeks, and data were assessed to find the conditions of maximum removal along with relevant isotherms and kinetic equations. The results displayed the soil samples followed the order sandstone > conglomerate > siltstone > sand for the removal of Ni (II), and Zn (II), and phenol from synthetic produced water for 250 mg/L for the soil samples. The sandstone displayed the highest removal of 98-99% while conglomerate and siltstone presented 89-98% for Ni (II) and Zn (II), and 32% for phenol, respectively. The experimental data were fitted using the Langmuir isotherm, Freundlich isotherm, Temkin isotherm model, and D-R isotherm, and the results described the best fit for the Langmuir isotherm for Ni (II) and Zn (II) and Freundlich isotherm for phenol. The maximum adsorption capacity was corresponding to complete single monolayer coverage from Langmuir isotherm of sandstone for Ni (II), and Zn (II) is 2.3 mg/g and 4.4 mg/g, respectively which is consistent with experimental values (2.6 mg/g for Ni (II) and 3.6 mg/g for Zn (II)). The equilibrium kinetic model studies were presented four samples that followed the pseudo-second-order kinetics for the removal of metal ions and phenol than pseudo-first-order kinetics and Elovich kinetics. The 2D correlation analysis combined with Fourier transform infrared spectroscopy in attenuated total reflectance geometry (2D COS FTIR) was applied to investigate the mechanism of removal of Ni (II) and Zn (II), and phenol. The 2D-COS FTIR analysis was verified that removal of metal ions/phenol can be driven by both silica and carbonate groups present in the minerals of four soil samples. The asymmetric Si-O as the prime response in Ni (II), and Zn (II) removal mechanism for sand and siltstone and asymmetric CO_3^{2-} bands decided for the sandstone and conglomerate. While the interaction of phenol and Si-O bands was predominant in the removal mechanisms for phenol along with electrostatic interaction between phenol and carboxylate in four soil samples. The characterization studies such as SEM, EDX, FTIR, and XRD analysis after the treatment of soil samples with produced water were indicated calcite minerals in sandstone and conglomerate, and silica-based minerals in siltstone and sand are the main responsible factor for the highest removal of Ni (II), Zn (II), and phenol which is consistent with results of 2D-COS FTIR analysis. However, the ICP-MS analysis of treated produced water presented that some components such as magnesium, calcium, and sodium from four soils were discharged into treated water even though high removal performance. The percentages of leached materials in treated water were different from each other because of the difference in the percentage composition of minerals present in each soil samples which was confirmed by XRF analysis. In summary, sandstone, conglomerate, and siltstone provided a high degree of contaminants removal with minimum treatment cost, thus these soils can be effectively used for SAT process for the treatment of produced water. Furthermore, the 2D-COS FTIR analysis offered a unique insight into a clear picture of the mechanism of removal on the surface of the soil by the interaction between surface functional groups containing Si-O bond, and CO_3^{2-} groups of soil samples and metal ions or phenol.

533s Joining Lab Experiments and CFD-Amozone Model for Deep Understanding of Batch Ozonation and Improve Experimental Method

Giacomo Bellandi, Roberta Muoio, Min Yang, Kevin Guerrero, Arantxa Sanchez, Andrea Börgers, Ruud Schemen, Peter van Dijk, Pieter Vlasschaert, Usman Rehman, Arne Wieland, Jochen Türk, Wim Audenaert, Ingmar Nopens

Introduction

Water treatment and (indirect) potable water reuse (IPR) is nowadays a topic of major attention due to the rising (and interconnected) concerns about water scarcity and micropollutants (MPs) such as pharmaceuticals and pesticides. Advanced tertiary treatments are increasingly adopted at water treatment sites to eliminate compounds with elevated biological resistance and produce safe water for human consumption (e.g. treatment to drinking water standards prior to aquifer discharge). In the US there are multiple examples (e.g. the HRSD SWIFT project (swiftva.com)) aimed at environmental protection from MPs. Conventional water resource recovery facilities (WRRFs) only partially contribute to these goals, and often in a rather unpredictable manner (Schaar et al., 2010).

Ozonation and its combination with other advanced oxidation processes (AOPs) is establishing more and more as an attractive IPR technology alternative for MP removal. Ozone has complex reactions with the effluent matrix. Some of those pathways lead to hydroxyl radical (HO*) formation. Both ozone and HO* will oxidise the MPs. However, depending on the natural Br- concentration in the water, the carcinogenic by-product bromate (BrO3) can be formed (Fischbacher et al. 2015). In addition to this, the specific ozone demand is strongly water-matrix-dependent, leading to significant differences in BrO3 formation, MP removal efficiencies, and ultimately operational costs.

Batch ozonation experiments have been used for decades as a method to evaluate the ozone demand of a given water matrix, on the suitability of ozone technology for micropollutants (MPs) removal, and even on its potential BrO_3 production. However, no standardized technique is currently available. Ozonation tests performed by different laboratories on the same water matrix show that the output of such experiments can be strongly biased by the way it is executed, especially when BrO_3 formation is a concern. The time and speed of injection, as well as the location where O_3 solution is supplied can drastically change the results. Even the O_3 stock solution concentration itself can influence the BrO_3 formation.

In this work, by coupling experimental tests and the use of the Amozone ozonation kinetic model integrated with computational fluid dynamics (CFD), important insights on the performance of batch ozonation tests were investigated for a secondary effluent matrix from the WRRF of Eindhoven (The Netherlands). The simulations in 3D and in millisecond time scale revealed the complex net of reactions involved in a batch experiment and how its execution can be better standardized to have more consistent results regarding the removal of MPs and the BrO3 formation and ensure data reproducibility.

Experiments were focused on testing the effect of ozone dose, stock solution concentration, and time of injection on MPs removal and BrO3 formation. The Amozone kinetic ozonation model was applied in CFD environment to replicate the experimental results and understand where the main improvements in the methodology were possible. The highest BrO_3 formation obtained in the batch ozonation tests was generally due to the higher local formation of OH* caused by either a faster injection or the use of a more diluted O_3 stock solution.

The joint effort between lab experiments and CFD simulations coupled with the Amozone kinetic ozonation model will contribute to set standards for the execution of batch ozonation tests and gain confidence in the applicability of ozone technology.

Methodological Approach

Batch ozonation tests

A total of 4 batch ozonation tests (Table 1) were planned in duplicates on a 24h composite water sample from the WRRF of Eindhoven (The Netherlands) in order to test the ultimate effect of the experimental protocol on the BrO_3 formation. The raw sample composition is reported in Table 2. In these tests O_3 residuals and HO* exposure (using pCBA probe component) were monitored at different moments in time with 10s time resolution. BrO_3 was measured both in the raw and ozonated sample.

Amozone *model combined with CFD*

Amozone is a kinetic model of ozonation and advanced oxidation processes that realistically predicts O_3 and HO* in real water matrices (Audenaert et al., 2019). The model describes the rapid and slow O_3 consumption, HO* production and scavenging, O_3 and HO* based MP oxidation, bromate formation mechanisms (von Gunten, 2003), nitrite quenching, chloramines formation and reactions, and several pH equilibria. This mechanistic model was embedded in the hydrodynamic simulation of the batch ozonation test replicating the experimental setup and its flow field in transient mode during the whole duration of each experiment. The kinetic model was calibrated using the experiment number 1 and validated by replicating the results of experiment 2 and 3.

An additional (virtual) test was performed using the same experimental conditions of test 3, removing NH_4^+ from the sample composition to appreciate the buffer effect of NH4-N on the BrO_3 formation.

Results and Discussion

The BrO_3 formation in the different experiments (Figure 1), despite the use of the same water matrix and laboratory tools, can significantly differ depending on the procedure used. The stock solution concentration had the strongest impact on BrO_3 formation due to the higher sample dilution with the addition of more stock solution to reach the same O_3 dose, e.g. in test 3 the same O_3 dose as test 1 was delivered by adding the double of the stock solution. The dilution effect lowered the OH* scavenging by the water matrix causing higher OH* exposure, as confirmed by the probe component pCBA (Figure 2). The time of injection also had a significant impact on the BrO_3 formation.

Combining the Amozone chemical kinetics with the hydraulic model in CFD was possible to understand how the several reactions with the water matrix involving O_3, OH*, and Br-intermediates evolved. During the first instants of the injection, reactions involving O_3 and organics occur, leading to consumption of O_3 and OH* formation. The presence and the ratio of O_3 and HO* decide the extend of the BrO_3 formation together with the Br^- availability.

Local high O_3 concentrations can be observed shortly after injection (Figure 3, top) while at the same time, the generation of OH* causes the depletion of the probe component pCBA (Figure 3, bottom). For the first few seconds after the injection, the O_3 stock solution spreads unevenly in the beaker from the injection point and a cloud of highly concentrated O_3 starts to react with the water matrix. The virtual experiments confirmed that the use of the less concentrated stock solution caused less HO* scavenging in experiment 3 as compared to experiment 1, and generated BrO_3 via the more efficient pathway through the BrO* disproportionation (data not shown).

Comparing the experiment 3 (model validation) with the results of a virtual test in absence of NH_4^+ (while maintaining the same experimental method and sample composition), higher BrO_3 (55 ug/L) would be formed in case the WRRF would have completely removed the NH_4^+.

Conclusions

Local O_3 and OH* concentrations are severely impacted by the way that ozone batch tests are performed, and they directly impact the final BrO_3 formation. This can lead to biased decisions when it comes to assessing the suitability of ozonation for a specific water. Therefore, a standardized method of performing ozone batch experiments need to include detailed guidelines on the speed of injection and stock solution dose. In addition to this, the magnetic stirrer speed could also have an important impact on the results. Further virtual tests will investigate mixing effect among other experimental conditions for different water qualities in order to provide solid ground for the standardization of batch ozonation tests.

The power of coupling experiments with virtual testing is threefold: i) investigate with unprecedented spatial and temporal detail chemical mechanisms which are water matrix-specific, ii) test conditions which are impossible to test in reality (e.g. removing NH_4^+ without altering the sample), and iii) replicate the same procedure across experiments maintaining the same water quality.

The joint effort between lab experiments and CFD simulations coupled with the Amozone kinetic ozonation model will contribute to set standards for the execution of batch ozonation tests and gain confidence in the applicability of ozone technology.

References

Audenaert, W., Bellandi, G., Pearce, R., Takács, I., Buehlmann, P., Hogard, S., Salazar-Benitez, G., Rehman, U., Nopens, I., Wilson, C., Bott, C.B., 2019. A novel kinetic ozonation model for prediction of bromate formation, bromate mitigation and trace organic contaminant removal, in: 24th World Congress of the International Ozone Association.

Fischbacher, A., Löppenberg, K., Von Sonntag, C., Schmidt, T.C., 2015. A New Reaction Pathway for Bromite to Bromate in the Ozonation of Bromide. Environ. Sci. Technol. 49, 11714–11720. https://doi.org/10.1021/acs.est.5b02634

Schaar, H., Clara, M., Gans, O., Kreuzinger, N., 2010. Micropollutant removal during biological wastewater treatment and a subsequent ozonation step. Environ. Pollut. 158, 1399–1404. https://doi.org/10.1016/j.envpol.2009.12.038

von Gunten, U., 2003. Ozonation of drinking water: Part I. Oxidation kinetics and product formation. Water Res. 37, 1443–1467.

533t Application of Waste Materials As Demulsifier for Oil/Water Separation

Ahmad Adewunmi, Muhammad Shahzad Kamal, Shirish Patil, Syed Hussain, Afeez Gbadamosi

During the extraction of crude oil, water production is a serious problem as most of the oil is produced in the form of stable emulsion due to the presence of natural emulsifier in the crude oil. These emulsions could be water-in-oil or oil-in-water depending on the interactions between the different components of the emulsion. Before transporting to refineries, the separation of water from the crude oil is extremely important as water can cause corrosion and other operational challenges during processing. Several mechanical and chemical methods are used to break the emulsion and separate the water from crude oil. Chemical demulsification is one of the important technique to separate the water by breaking the emulsion. Several chemicals such as ionic liquids, nanoparticles, surfactants, and polymers are used as demulsifier. A significant work has been published on the application of these materials in chemical demulsfication. In this work, we propose the use of natural material as demulsifier to break the highly stable emulsion. Several demulsifiers such as quartz obtained from sand, palm oil fuel ash, and fly ash was applied to break the high stable emulsion. The emulsions were prepared using water and different types of crude oils using a high speed mixer. The emulsions were prepared by varying the concentration of salts and water/oil ratio to mimic the real conditions. The asphaltene present in the crude oil acts as natural emulsifier that stabilize such emulsion. Demulsifiers were added to these emulsions to evaluate the performance of these demulsifiers. A bottle test was used to measure demulsfication efficiency (DE%). Demulsification tests were done at room temperature and at high temperature. Emulsion characterization was done using droplet size measurements, interfacial tension measurement, and rheological measurement. The demulsfication bottle test indicates that the reference emulsion without any demulsifier was stable for several days and no separation of water was observed. However, the addition of different natural demulsifier at different concentrations break the emulsion and separated the water from the oil. Using these low cost demulsifier resulted in breaking the emulsion with different DE depending on the temperature, type and concentration of salts, dosage of demulsifier, and type of the demulsifier. More than 97% demulsification efficiency could be achieved using these low cost materials.

533u Hydrogen Production Via Ammonia Decomposition over Ni-Pt/Al_2O_3 using Electrical Heater: A Multi-Objective Optimization

Ali Cherif, Chul-Jin Lee

Process electrification is a viable solution to override the perpetual dependency on the depleting carbon-based fossil fuel. In this study, hydrogen production via electrically heated ammonia decomposition reaction over Ni-Pt/Al2O3 catalyst was numerically modeled and analyzed. In this study, the effect of the heater configuration on the reactor performance was analyzed adopting 15 cases. Furthermore, a multi-objective optimization approach of the heater design was applied to achieve the highest ammonia conversion and the lowest heat release from the heater via finding the optimal heater configuration and the proper inlet temperature and velocity. The results showed improved ammonia conversion in most of the suggested cases compared to the conventional heater configuration with the highest conversion equal to 95.97%, compared to the conventional heater, however, with a trade-off in the consumed heat (i.e., 2805.04 W). The sensitivity analysis showed a high impact of the heater configuration compared to the velocity and temperature. The multi objective optimization showed an improvement of 4% in the ammonia conversion, and a reduction of 45% in the supplied heat from the heater under an inlet temperature of 897 K and inlet velocity of 0.52 m/s. The proposed optimal configuration might be a reasonable solution for the economic feasibility of the electrically heated ammonia decomposition.

533w Methanotrophic Activity in the Deep Environment: Enhancement of Methane Catalysis Rates

Dipayan Samanta, Rajesh K. Sani

Methane is a powerful greenhouse gas, second to carbon dioxide in abundance and yet over 25-times as effective in trapping heat. The impact of methane fluxes from the deep biosphere of non-coal mines on microorganisms and the associated microbiome involved in methane oxidation has not been previously studied. There are still several intriguing questions about methanotrophs and their key methane monooxygenase (MMO) enzymes. MMO in methanotrophs are uniquely responsible for the oxidative catalysis of an important biochemical process that is the conversion of methane to methanol. There are two forms of MMO that exists in methanotrophs- cytoplasmic or soluble methane monooxygenase (sMMO) and transmembrane or particulate methane monooxygenase (pMMO). However, there is a controversy regarding the active site(s) in pMMO, specifically whether the active site(s) are in the *pmoB* (α subunit), *pmoA* (β subunit), or *pmoC* (γ subunit). Also, MMOs have lower affinity towards methane.

With our years of exploration in the deep biosphere (300 to 5,000 ft. levels) of the Sanford Underground Research Facility (SURF) at Homestake Gold Mine (Lead, SD, USA), we have confirmed the synergy among more than 44 different bacterial and archaea phyla in water, soil, sediments, and rock samples. Our results indicated the domination of *Rhodobacteraceae, Alcaligenaceae, Bradyrhizobiaceae, Mycobacteriaceae, and Pseudonocardiaceae* families in deep biosphere of SURF. Certain microbial members of these families had sMMO. The relative importance of *Rhodobacteraceae* in SURF and our results with *Rhodobacter* sp. showing their ability to carry out methanotrophy suggest that evolutionary groups other than conventional methanotrophs are involved in methane oxidation. These findings suggest that there may be more novel methanotrophs that are unidentified and virtually unstudied.

Using systems biology approaches, we unveiled whether the active site(s) of pMMO is located within the copper center or within the vicinity of the copper center using an obligatory aerobic methanotroph *Methylosinus trichosporium* OB3b as a model organism. To increase the catalysis rates of methane in pMMO of OB3b, selected amino acid residues interacting at the binding site of ethylbenzene, toluene, 1,3-dibutadiene, and trichloroethylene were mutated. Based on screening the strain energy, docking energy, and physiochemical properties, five mutants were down selected, B:Leu31Ser, B:Phe96Gly, B:Phe92Thr, B:Trp106Ala and B:Tyr110Phe, which showed docking energies of -6.3, -6.7, -6.3, -6.5 and -6.5 kcal/mol, respectively as compared to the wild type (-5.2 kcal/mol) with ethylbenzene. These results suggest that these five mutants would likely increase methane oxidation rates compared to the wild type pMMO.

Furthermore, in our wet lab experiment, we did take into consideration the physiochemical method of enhancement of methane catalysis rate in *Methylosinus trichosporium* OB3b using supplementation of copper. A decade of methanotrophic research witnessed the significance of copper as a crucial metal in the differential expression of sMMO and particulate pMMO forms of MMOs. In certain methanotrophs, expressing both forms of MMOs, where the expression of either form is usually controlled by the concentration of Cu, following a mechanism referred as the "Cu switch". The switch has an enthralling role play during the bioremediation of effluents as sMMO and pMMO degrades broad range organic substrates but diverges in specificity. In our study, *Methylosinus trichosporium* OB3b was used as a model bacterium to investigate the range of Cu concentration that triggers the expression of sMMO to pMMO, the ensuing effect of altering Cu concentration on the growth, methane oxidation capabilities, and the expression of individual genes within the methanobactin operon of OB3b. The Cu switch was observed to be regulated within 3 μM to 5 μM of Cu concentration with a strict increase in the methane consumption rates per day from 49 to 78 mg/L (at the mid exponential phase). The consumption of oxygen was observed to be proportional to the methane utilization rate and was highly controlled by the di-heme enzyme within the methanobactin operon. The ICP-MS data showed that Cu has been utilized at a greater rate when more Cu was supplemented in the media, and simultaneously the upregulation of methanobactin synthesis genes (mbnABC, and mbnT) signified that more Cu was harvested when pMMO genes were upregulated (≥ 5μM). A newly developed quantitative assay based on Naphthalene-Molisch principle were performed to distinguish between the sMMO and pMMO expressing cells. The gene expression results coincided with the assay where naphthol concentration was detected to be higher at low Cu concentration, and gradually decreased to no naphthol production where pMMO was expressed and acted as the sole methane oxidizer. However, the trend of increase in methane oxidation became inconsistent when the cells reached its toxicity level (40 μM). Therefore, to elucidate the mechanism and the involvement of crucial genes towards copper toxicity and copper homeostasis, more proteins were considered using physical protein-protein interaction. This resulted in 7 transporter protein, 3 cell wall biosynthesis/degradation protein, a Cu resistance operon (*cop*), and 20 hypothetical proteins may also be involved in the Cu response to the environment and Cu regulation within the organism.

533y Understanding the Transport and Fate of the Microplastic Particles in a Main Wastewater Treatment Plant: Case Study of a Typical Plant in Western Iran

Shafieh Karami, Rashed Rashidi, Pedro E. Arce, Negar Saraei

Large quantities of microplastics (MPs) are thought to be globally discharged to the freshwater environments through wastewater treatment plants (WWTPs). However, the information available related to sources of contamination is limited. This case study focusses on understanding the role of water effluent and sludge from a municipal WWTP (located in Western Iran) as potential pathways for MPs input into the both aquatic and agriculture-based environments. The Sanandaj's WWTP has a large daily treatment capacity and serves about 500,000 residents. Samples from different plant locations were taken: For example, from influent, grit removal chamber outflow, primary effluent, and secondary effluent. Sludge samples were also collected and investigated to evaluate the potential risk associated with the application of dried sewage sludge as fertilizer in agricultural soils. Samples were filtered through a set of sieves in laboratory and, afterward, they were counted and classified according to their shape and size by means of visual investigation using a stereomicroscope. The results showed that the most dominant particles were fibers followed by fragments. Furthermore, Fourier-transform infrared spectroscopy (FTIR) was used to confirm that the plastic originated from suspected particles. The MPs concentration ranged from 163 MPs/L in influent, that measures up to an annual arrival of 595×10^{10} MPs into the plant, to 8.4 MPs/L and 96 MPs/g in effluent and dried sludge, respectively. This data indicates that even though the MPs separation efficiency of the plant was high (>90%), approximately 30.7×10^{10} and 67.3×10^{10} MPs are released into the receiving environment each year by dissemination of final effluent and dried sludge, respectively. One key outcome of the current case study is confirming the presence of MPs in WWTP (such as Sanandaj's) effluent and sewage sludge. Moreover, the study indicates that WWTPs could be an important pathway by which MPs are introduced to the marine and terrestrial environments. The present case can be duplicated in other WWTP in different locations and validate the trend so that treatment procedures can be identified for minimizing or preventing the discharge of MP to water streams.

629b Chemical Recycling of Agricultural Waste Plastic By Pyrolysis: Case of Study of Province of Buenos Aires, Argentina

Gabriel Rodriguez Garrido, Pablo Marinangeli, Maria Alicia Volpe, Patricia Hoch, Maria Soledad Diaz

This study addresses the use of pyrolysis as a way to carry out the chemical recycling of agricultural waste plastic (AWP) in the southwest region of Buenos Aires province in Argentina. Large containers of agrochemicals and silo bags are the main sources of AWP in this area of the country, and pyrolysis has been found as one acceptable treatment, as it allows obtaining energy, fuels or products of better added value than simply incinerating the residues. Pyrolysis also allows to reduce the carbon footprint when compared to other thermal treatments commonly used to dispose this kind of waste.

The liquid and gas products obtained by pyrolysis of AWP are characterized in the context of producing electric and thermal energy. Yields of liquid and gas productsa are 40 and 48% respectively. Besides, the solid co-product is studied.

The products, as well as the solid co-product, are characterized by thermal analysis (DSC/TGA), elemental analysis, lower heating value (LHV) measurements, X-ray diffraction and gas chromatographic analysis. Results indicate that gas and liquid show high LHV (49,6 and 46, 9 MJ/Kg respectively). Pyrolysis technology is compared to both mechanical recycling and incineration of plastic, considering CO_2 emissions. In addition, the energetic efficiency of the pyrolysis process of AWP is calculated, considering the liquid and gas products as alternatives to fuel-oil and natural gas respectively.

Conclusions

Pyrolysis of AWP has a significantly lower carbon footprint compared to incineration, while the process is adequate to recycle materials inapplicable for conventional recycling. The liquid is also considered as a potential material in the synthesis of circular economy plastics.

473e Advanced Oxidative Degradation of Polyethylene for Sustainable Upcycling

Shoumik Sadaf, Zhihua Jiang

Plastic has been a global environmental issue due to its slow degradation rate in the environment and its negative impacts towards soil and aquatic systems. The objective of this research is to establish an innovative technology for sustainable upcycling of polyethylene, the most common plastic in use today. Our approach to achieve this objective is to systemically investigate the combination of two of the most promising advanced oxidation techniques: Photocatalytic oxidation and ozonation to degrade polyethylene into fragments with different chemical functional groups that can be used as feedstock for filamentous fungi to produce high value products. For enhancing the degradation, zinc oxide nanopowder was coated with graphene and was used as photocatalyst. The preliminary analysis showed varied degrees of weight degradation of pure low density polyethylene with the formation of carbon oxygen double bonds. We will optimize the advanced oxidation conditions and systematically characterize the oxidation products using various analytical techniques including GPC (Gel Permeation Chromatography), GCMS (Gas Chromatography Mass Spectrometry), HPLC (High-performance liquid chromatography) and NMR (Nuclear Magnetic Resonance) Spectroscopy.

57a The Breathing Human Infrastructure: Integrating Air Quality, Traffic, and Social Media Indicators

Marwa El-Sayed, Heather O'Leary, Scott Parr

Outdoor air pollution is a complex system that is responsible for the deaths of millions of people annually, yet the integration of interdisciplinary data necessary to assess air quality's multiple metrics is still lacking. This case study integrates atmospheric indicators (concentrations of criteria pollutants including particulate matter and gaseous pollutants), traffic indicators (permanent traffic monitoring station data), and social indicators (community responses in Twitter archives) representing the interplay of the three critical pillars of the United Nations' Triple Bottom Line: environment, economy, and society. During the watershed moment of the COVID-19 pandemic lockdowns in Florida, urban centers demonstrated the gaps and opportunities for understanding the relationships, through correlations rather than causations, between urban air quality, traffic emissions, and public perceptions. The relationship between the perception and the traffic variables were strongly correlated, however no correlation was observed between the perception and actual air quality indicators, except for NO_2. These observations might consequently infer that traffic serves as people's proxy for air quality, regardless of actual air quality, suggesting that social media messaging around asthma may be a way to monitor traffic patterns in areas where no infrastructure currently exists or is prohibited to build. It also indicates that people are less likely to be reliable sensors to accurately measure air quality due to bias in their observations of traffic volume and/or confirmation biases in broader social discourse. Results presented herein are of significance in demonstrating the capacity for interdisciplinary studies to consider the predictive capacities of social media and air pollution, its use as both lever and indicator of public support for air quality legislation and clean-air transitions, and its ability to overcome limitations of surface monitoring stations.

57b Assessing Vehicle Contribution for CO and CO2 Emission in Los Angels Using BECO2N Sensor Network

Suverna Trivedi

In the United States, transportation accounts for the majority of anthropogenic CO and CO2 emissions in metropolitan areas. Urban transportation emissions must be reduced as soon as possible to meet climate goals set by international treaties, national policies, and municipal governments. Transportation emissions continue to be one of the most significant drivers to poor air quality (AQ) and AQ inequity. The Berkeley Air Quality and CO2 Network (BEACO2N) is collecting data of emissions in the Los Angeles area. We used BECO2N and PEMS data to determine the contribution of automobiles in the LA region near the BECO2N node in this study. In comparison to the multiple linear regression method, we recommend using a sequential neural network to learn about the contribution of the vehicles sector. We found that meteorological characteristics have the least impact on CO and CO2 emissions in the Los Angeles area.

57c Understanding the Performance of Low-Cost Sensors in Characterizing Particulate Matter (PM) in Outdoor and Indoor Environments

Zahra Shivji, Sabrina Westgate, Nga Lee Ng

With the progressive decrease in the cost of air quality sensors over the past few years, the potential applications of low-cost sensors for wide-spread air quality monitoring have grown. The rise of COVID-19 has particularly emphasised the application of using low-cost sensors to continuously monitor indoor air. However, the increase in low-cost sensor availability has also highlighted the importance of characterizing and quantifying low-cost sensor measurement capabilities. Towards this end, in this study we compared the particulate matter (PM) data reported by two low-cost sensors (Plantower PMS and QuantAQ MODULAIR-PM) in both outdoor and indoor environments on the Georgia Tech Campus. The Plantower sensor consists of a nephelometer and reports PM_1, $PM_{2.5}$, and PM_{10}, while the QuantAQ sensor reports size-resolved particle number concentration data from an AlphaSense optical particle counter (OPC) as well as mass concentration data (PM_1, $PM_{2.5}$, and PM_{10}) that is determined from a mathematical combination of nephelometer and OPC data. To investigate the effects of aerosol composition and size on the performance of the sensors, sensor data collected outdoors (urban Atlanta) in Winter 2021 were compared to measurements made by two research-grade instruments: Aerosol Chemical Speciation Monitor (ACSM) and Scanning Mobility Particle Sizer (SMPS). To assess sensor performance in indoor environments with PM emitted from various human activities, low-cost sensor data from the same time period were analyzed for three different indoor environments including a classroom, an academic common space, and a cafeteria. To further constrain sensor performance indoors, collocated sensor and SMPS measurements were performed in the classroom in Spring 2022. Results indicated that the data from both sensors generally agreed with research-grade instruments for PM_1 for both indoor and outdoor measurements. However, changes in aerosol mass loading and aerosol size distribution were observed to impact sensor performance and agreement. Specifically, during time periods with higher proportions of particles with diameters greater than 1 mm, the Plantower sensor predicted lower aerosol mass concentrations than the QuantAQ sensor. Results from this study provide data-driven insights into what types of indoor environments each sensor type is best suited for, and under what aerosol conditions they face the most limitations.

57d Enhanced Light Absorption and Radiative Forcing By Black Carbon Agglomerates

Georgios Kelesidis, David Neubauer, Liang-Shih Fan, Ulrike Lohmann, Sotiris E. Pratsinis

The climate models of the Intergovernmental Panel on Climate Change (IPCC) list CO_2, CH_4 and black carbon (BC), as the most potent contributors to global warming based on their radiative forcing (*RF*) impact. Among them, the BC contribution comes with the highest uncertainty (~ 90 %). Examining closely these models, it becomes apparent that they might underpredict significantly the direct *RF* for BC, largely due to their assumed spherical BC morphology (Kelesidis andamp; Pratsinis, 2021). In specific, the light absorption and direct *RF* of BC agglomerates are enhanced by light scattering between their constituent primary particles (Kelesidis et al., 2020) as determined by the Rayleigh-Debye-Gans theory interfaced with discrete dipole approximation (Kelesidis andamp; Pratsinis, 2019) and recent relations for the refractive index and lensing effect. The resulting light absorption agree very well with the observed absorption aerosol optical depth of BC. ECHAM-HAM simulations accounting for the realistic BC morphology and its coatings reveal high direct $RF = 3 - 5$ W/m^2 in East, South Asia, sub-Sahara, western Africa and the Arabian peninsula (Fig. 1). These are in agreement with satellite and AERONET observations of *RF* and indicate a regional climate warming contribution by 2.4 - 4 ^{o}C, solely due to BC emissions.

References:

Kelesidis, G. A., and Pratsinis, S. E. 2019. Soot light absorption and refractive index during agglomeration and surface growth. Proc. Combust. Inst. 37:1177-1184, doi.org/10.1016/j.proci.2018.08.025.

Kelesidis, G.A., Kholghy, M.R., Zuercher, J., Robertz, J., Allemann, M., Duric, A., and Pratsinis, S.E. 2020. Light scattering from nanoparticle agglomerates. Powder Technol. 365:52–59, doi.org/10.1016/j.powtec.2019.02.003.

Kelesidis, G. A., and Pratsinis, S. E. 2021. Determination of the volume fraction of soot accounting for its composition and morphology. Proc. Combust. Inst. 38:1189-1196, doi.org/10.1016/j.proci.2020.07.055.

Figure 1. Global map of the direct *RF* of coated BC agglomerates estimated by ECHAM-HAM.

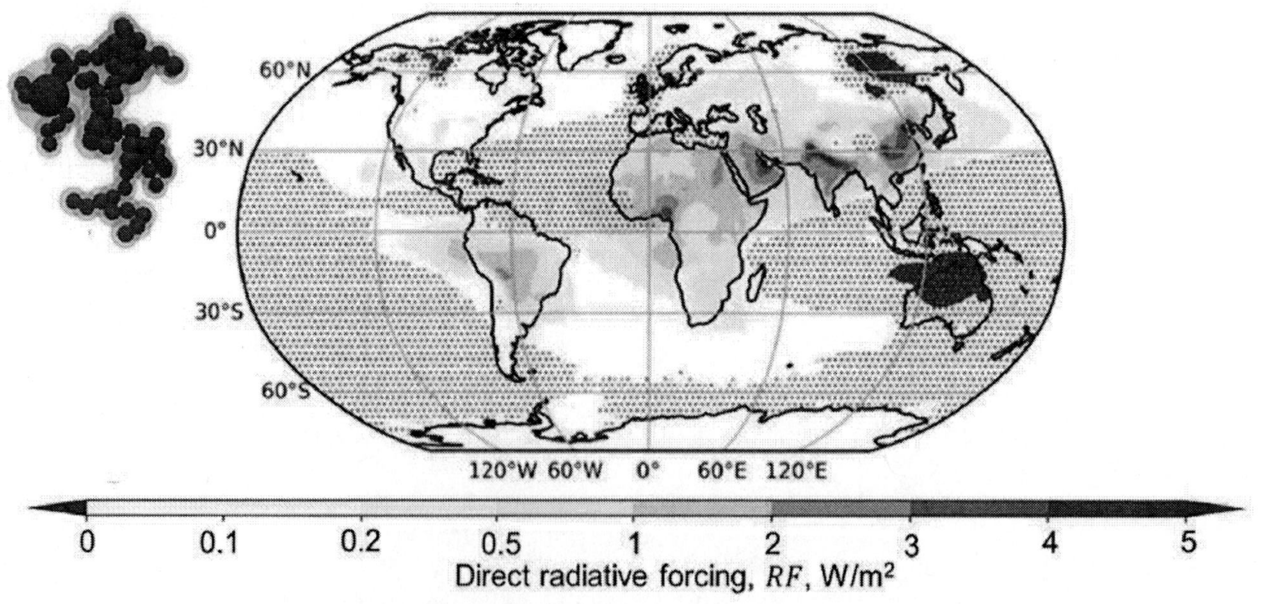

57e Agent Based Approach for Prescriptive Mitigation of Exposure to Air Pollution in Urban Setting

Pranav Agrawal, Sathish Swaminathan, Raghunathan Rengaswamy

Ambient air pollution has a severe effect on human health. Quantifying exposure to individuals is the first step in developing mitigation strategies which in turn depend on air quality data that is available with high spatio-temporal resolution. Existing infrastructure for monitoring air quality that uses static monitors is sparsely distributed and prohibitively expensive to expand. This results in gaps in capturing variability in air quality. Dispersion and CFD based models have been used extensively to fill such gaps. However these models are computationally intensive. Alternatively, hyperlocal air quality information obtained using mobile monitoring is considered in tandem with an agent based modeling framework. In this framework, a system is represented as a collection of agents interacting with each other and with the environment. At the heart of this framework is a mass balance equation, which determines the exchange of mass between the agents and the environment. The model captures high spatio-temporal variation in the air quality. Unlike dispersion and CFD based models which require complex differential-algebraic equations, the agent based framework takes a rule based approach, thus making it computationally less demanding. It also enables the model to capture emergent behavior in the system. In this work, we explore the applications of an agent based framework in analysis and prescription of mitigation strategies in an urban set up. The following applications are studied in detail under the framework:

1. Estimating exposure of an individual to air pollution as a function of their location and time activity information.
2. Determining the route between any two locations in city with the least exposure.
3. Dynamic traffic rerouting

A rule based approach makes it feasible to make such complex decisions for large urban road networks. The results are validated against field data.

194a Effect of Combustion Particle Morphology on Biological Responses in a Co-Culture of Human Lung Epithelial and Macrophage-like Cells

Kerry Kelly, Kamaljeet Kaur, Raziye Mohammadpour, Hamid Ghandehari, Robert Paine, Christopher Reilly

Background. Freshly generated combustion particles are hydrophobic and form fiber-like agglomerates with an average fractal dimension of 1.8. However, after aging in the atmosphere, particles become more hydrophilic allowing adsorption of water. When this water evaporates,it exerts capillary forces between the primary particles of the agglomerate, causing the structure to collapse and become more compact spheres-like. Previous studies with fresh and aged combustion particles have linked the changing chemical composition to be the prime suspect behind the observed differences in toxicological response. However, little is known about the contribution of morphological differences in atmospherically aged particles to their toxicological response, possibly due to the difficulty in resolving the two properties (composition and morphology) that change simultaneously. A method to change particle morphology without the need to change chemical composition would help elucidate the effect of this morphological change on cellular responses.

Methods. This study altered the shape of lab-generated combustion particles, without affecting the chemical composition, from fiber-like to a more compact spherical shape using a water condensation-evaporation method. Using an electrostatic field-based air-liquid exposure (ALI) chamber, the two shapes were exposed to a co-culture of human airway epithelial (A549) and differentiated human monocyte (THP-1) cells at ALI conditions.

Results. For the same mass dose of 2 mg/cm^2, both shapes were ingested by cells, and induced a pro-inflammatory response (IL-8 and TNFα) and enhanced CYP1A1 gene expression compared to air controls. The more compact spherical particles (representative of atmospherically aged particles) induced more early apoptosis and release of TNFα compared to the more fiber-like particles.

Conclusions. The results suggest a contribution of morphology to the increased toxicity of aged particles.

194b Using Low-Cost Air Purification to Mitigate COVID-19 Transmission in Schools and Community Settings

William Gasparrini, Kristina Wagstrom

As COVID-19 continues to spread globally and societies continue to lift many of the mitigation effort restrictions, there is a critical need for effective, accessible approaches to filter indoor air to mitigate COVID-19 transmission. Conway-Morris et al.(2021) demonstrated that commercial HEPA air purifiers can effectively remove the size aerosols most associated with transmission of COVID-19, but these units can be cost prohibitive. Lower cost, commercially available air purifiers do not effectively remove the aerosols most responsible for the transmission of COVID-19. The Corsi-Rosenthal box filter is a DIY (do-it-yourself) solution that has the potential to fill this gap as it is cost-effective (~$75), energy efficient, and able to remove the aerosol size most associated with transmitting COVID-19. Lab-based analyses of Corsi-Rosenthal boxes have demonstrated their effectiveness.

A team comprised of researchers and clinicians from UConn's School of Engineering and School of Nursing and UConn Health's School of Medicine have deployed Corsi-Rosenthal boxes throughout school and community spaces in Connecticut. Volunteers built these boxes during several volunteer build events, and first-year engineering students built 100 as part of their freshman engineering course. We will present the findings from (1) real world testing of the Corsi-Rosenthal boxes in active classroom settings, (2) modeling of the air circulation when using two Corsi-Rosenthal boxes on low speed, (3) analysis of the effectiveness of the course-based assignments for helping student achieve the course objectives, and (4) a summary of the community perception of the Corsi-Rosenthal boxes deployed in different indoor educational settings in Connecticut.

194c Hybrid Water Activity – a Novel Framework for CCN Analysis of Sparingly Water Soluble Organic Aerosols

Kanishk Gohil, Chun-Ning Mao, Dewansh Rostogi, Chao Peng, Mingjin Tang, Akua Asa-Awuku

Organic aerosols are comprised of hundreds of organic compounds either emitted directly into the atmosphere (Primary Organic Aerosols; POAs) or formed due to chemical transformations in the atmosphere (Secondary Organic Aerosols; SOAs). One such class of compounds are aromatic acids that can act as both POAs and SOAs. Phthalic acid (Pth), isophthalic acid (IPth) and terephthalic acid (TPth) are some of the significant benzene polycarboxylic acids. The molecular structure of Pth, IPth and TPth contains two carboxylic acid groups attached on a benzene ring. Pth, IPth and TPth are structural isomers and differ significantly from each other in terms of water solubility which affects their cloud condensation nuclei (CCN) activity. So far, the CCN activity of Pth and IPth has been studied and explained using the traditional Köhler Theory (KT). The CCN activity and hygroscopicity of TPth has not been characterized in the past in the context of cloud activity and droplet formation. Moreover, KT can be effectively applied for CCN activity analysis of Pth due to its higher aqueous solubility compared to IPth and TPth. In this work, we studied the CCN activity and droplet growth for Pth, IPth and TPth. We developed the Hybrid Activity Model (HAM) by incorporating the aqueous solubility of a solute within an adsorption-based activation framework. Frenkel-Halsey-Hill (FHH)-Adsorption Theory (FHH-AT) was combined with the compound solubility to develop HAM. Analysis from HAM was validated using laboratory measurements for pure Pth, IPth, TPth and Pth-IPth internal mixtures. Furthermore, the analysis results generated using HAM were tested against traditional KT and FHH-AT to compare their water uptake predictive capabilities for the compounds selected for this work. Additionally, a single-hygroscopicity parameter was also parameterized based on the FHH-AT and HAM frameworks. The hygroscopicity parameter developed using FHH-AT provided a better quantification of IPth and TPth CCN activity as compared to the hygroscopicity parameter based on KT. Moreover, an improved agreement in experimental FHH-AT hygroscopicity was observed with respect to theoretical FHH-AT hygroscopicity for the AAC-CCN measurements. It was also found that the hygroscopicity parameter based on HAM could successfully simulate the water uptake behavior of the pure and internally mixed samples. These computations were found to be in line with either traditional KT or FHH-AT depending on the aqueous solubility of the sample.

194d Combined Ice Crystal Growth Chamber and Flow System to Investigate Atmospheric Secondary Ice Nucleation

Sylvia Sullivan

At temperatures warmer than -40°C in the atmosphere, ice crystals can nucleate heterogeneously on a subset of aerosol particles, but observed ice crystal number concentrations often exceed corresponding ice-nucleating particle concentrations. An array of *secondary production processes* have been proposed to explain this discrepancy, including crystal breakup upon collision analogous to impact attrition within industrial crystallizers. We present a two-component lab setup to study these processes through the lens of secondary nucleation during crystallization. First, pristine ice crystals are grown in an air-filled nucleation chamber using distinct supersaturation and temperature histories to generate distinct habits. We characterize the temperature and supersaturation gradients and precisions within this chamber, as well as the array of resultant crystal geometries. Seed crystals are then transferred to a flow system where they are mounted atop a thermistor and contacting rod to perform collisions which are photographed with a high-speed, high-resolution camera. We show preliminary results of impact energy dependence in the ice crystal number generated by these collisions.

56a Photocatalytic ZnO Foams for Organic Micropollutant Degradation

Davide Mattia

The presence of micropollutants - pharmaceuticals, pesticides, phthalates and hormones - in water is a major global health and environmental challenge. Conventional wastewater processes are not capable of effectively removing these compounds, often found at very low concentrations, $\mu g\ L^{-1}$ or even ng L^{-1}, leading to their discharge in surface and ground water. Advanced oxidation processes (AOPs) while effective at removing micropollutants, have not yet been deployed at industrial scale due to cost and safety concerns associated with the potential leaching of nanoparticle slurries on one hand, and the low activity of immobilised photocatalyst configurations on the other.

Here we report on novel photocatalytic foams, whose activity is given by the material itself rather than grafted nanomaterials, as previously done [1]. ZnO photocatalytic foams were prepared using a two-step procedure of liquid templating and sintering: Air was first incorporated into a dispersion of ZnO particles, surfactant and polymer. Upon crosslinking reaction and drying, a highly porous self-supporting foam, composed of pure oxide particles sintered together into an interconnected monolith, was produced. Morphological and structural characterisation was performed using FE-SEM, Raman, XRD and microcomputed tomography. Degradation experiments were performed using carbamazepine (10 $\mu mol\ L^{-1}$) as a model pollutant in both recirculating and single-pass reactors under UV irradiation with flow rates ranging from 100 to 500 mL min^{-1} and 0.7 to 5.0 mL min^{-1}, respectively, under oxygen saturation and adsorption-desorption equilibrium time of 40 min.

Under recirculation conditions at 500 mL min^{-1}, 76% degradation of carbamazepine was obtained after 2 h compared to 4 h for a batch reactor, with a quantum yield of 0.60 and electrical energy per order below 1.0 kWh m^{-3}. At lower flow rates, the effect of mass transfer resistances was significant, with a reduced degradation performance. The foams showed high chemical and mechanical stability, with minimal Zn leaching and reuse for 5 cycles and 10 hours of operation in total. Increasing the length of the foams from 2.1 to 6.3 cm led to a further increase in degradation of carbamazepine to 95% under the same conditions.

Under single -pass conditions, full degradation of carbamazepine was obtained in under 30 minutes for the lowest flowrate due to a combination of both photolysis and photocatalysis, with an enhanced quantum yield of 0.9.

These findings show that photocatalytic foams have comparable photocatalytic activity to and higher quantum yield and energy efficiency of nanoparticle slurries without the risks associated with nanoparticle leaching into the environment, which necessitates costly downstream removal. These results open the way to the safe industrial use of photocatalysis for the removal of micropollutants from water.

References:

T. Tasso Guaraldo, J. Wenk, and D. Mattia, 'Photocatalytic ZnO Foams for Micropollutant Degradation,' *Advanced Sustainable Systems*, p. 2000208, 2021, doi: https://doi.org/10.1002/adsu.202000208.

56b Continuous Flow Catalytic Hydrogenation of Groundwater Nitrate in a Membrane Catalyst-Film Reactor

Juliana Levi, Hunter P. Jacobs, Chen Zhou, Michael S. Wong, Bruce Rittmann, Paul Westerhoff

Nitrate is the most commonly occurring anthropogenic groundwater pollutant in the US, due largely to over application of nitrogen-based fertilizers. In drinking water, nitrate consumption can cause oxygen deficiency in the bloodstream; an especially dangerous occurrence for infants. Accordingly, the US EPA has imposed an MCL of 10 mg NO_3-N/L. Existing technologies to treat nitrate focus on separating nitrate from water using ion exchange or reverse osmosis that produce brine waste streams that require disposal. Also, a few full-scale treatment systems currently operate using biological denitrification for drinking water treatment wherein nitrate is converted to innocuous by-products.

Non-biological catalytic hydrogenation rapidly converts nitrate to nitrogen gas using hydrogen gas (H_2) and a solid-phase catalyst. Unlike biological systems, these abiotic catalytic hydrogenation can be turned on or off with near zero start-up time. However, 80% of publications regarding catalytic hydrogenation treatment option focus on material development and employ batch slurry reactor with a catalyst in suspension and uncontrolled hydrogen flow, which is not scalable for drinking water treatment.

The presented technology uses hollow fibers as a method of controlling hydrogen delivery in a continuous flow reactor. This method improves the efficiency of hydrogen delivered, which results in an order of magnitude decrease in required hydrogen flow rate and cost. Previously studied nitrate reduction systems have incorporated catalysts in suspension, but those configurations risk metal nanoparticles carrying over into drinking water. To resolve this technology barrier, the hollow fibers also act as a surface attachment point for the catalyst. After analyzing various catalyst coating methods, we have found a method to irreversibly attach nano-catalysts to the H_2-permeable hollow-fibers. The research focuses on a bimetallic Palladium-Indium catalyst because it has the highest selective reduction of nitrate to nitrogen gas without leaching. The parameters of the synthesized Pd-In catalyst and the reaction result in up to four times the nitrate reduction activity as similar catalysts while still achieving nitrogen selectivity above 95%. Reactor parameters and the corresponding effects on activity and selectivity of the catalyst will be discussed; particularly relating to scale up from batch to continuous flow. The reactor system is operated for extended periods to determine any sources of long term catalyst deactivation. This presentation focuses on the development of a scalable technology for groundwater nitrate treatment with high selectivity to desired nitrogen gas and minimized catalyst deactivation over time.

56c Inducing Nanobubble Collapse Via Dynamic Stimuli to Generate ·OH for Contaminant Degradation

Gregg P. Kotchey, David V. P. Sanchez

Due to their enhanced lifetime and stability, nanobubbles (NBs)—gaseous particles with diameters of 100 to 200 nm—have captured the attention of researchers for applications involving lake and pond remediation, horticulture, aquaculture, dissolved air flotation, scouring biofilms, and medical imaging. The enhanced properties afforded by NBs versus larger bubbles arise from: (1) movement of NBs is governed by Brownian motion (versus the buoyancy force that impacts the lifetime of larger bubble species); and (2) coalescence is hindered due to electrostatic repulsion between NBs, which have a negative zeta-potential.

Because of their fast reaction kinetics, strong oxidation potential, and ability to react with and mineralize organic contaminates, hydroxyl radicals (˙OH) represent a promising green oxidant for the removal of organic contaminants of emerging concern (CECs), pharmaceutical and personal care products (PPCPs), dyes, explosives, bacteria, and viruses during water and wastewater treatment. To date, the majority of literature on NBs contends that collapsing NBs would have insufficient energy to generate ˙OH. To realize the full potential of NBs derived from stable and readily available gas species as a source of ˙OH, NBs must undergo coalescence and subsequently collapse.

By altering the chemical conditions (i.e., ionic strength, pH, and gas species) as well as applying a dynamic stimulus (e.g., potential), ˙OH is produced in a controllable manner via the coalescence/collapse mechanism. For example, after the addition of 100 mM NaCl to air NBs, the relative size distribution determined by dynamic light scattering (DLS) indicated NB coalescence as evidenced by a change from a unimodal relative size distribution (black, mean diameter of 120 nm) to a bimodal distribution (red, mean diameters of around 100 nm and 450 nm). Meanwhile, Figure 1b depicts the synergetic effects of combining 100 mM NaCl with an applied potential (±250 mV, 5 min). While there is a minimal difference between the NB sample and DI samples after the application of +250 mV, the fluorometric tracer for ˙OH demonstrated a twofold increase for the NB sample versus the DI control after the application of-250 mV. The elucidation of the underlying conditions and mechanisms that govern the controlled coalescence and collapse will be critical for realizing the full potential of NBs as a source of ˙OH for sustainable environmental remediation efforts in the areas of wastewater and water treatment.

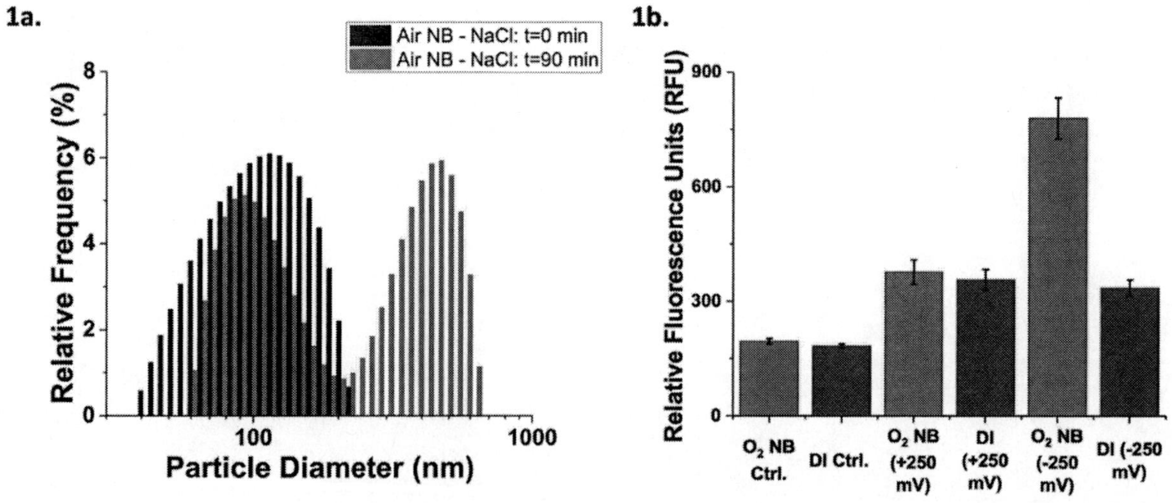

56d Chemical-Free Water Treatment with Immobilized Dual-Porous Photocatalyst

Mary Worbington, Ella Sheets, Sam Snow, Kevin McPeak

Advanced oxidative processes (AOPs), can be highly effective against a broad range of waterborne contaminants, but often require chemical additives (e.g. chlorine, peroxide, ozone) which increase costs and may lead to the formation of undesirable disinfection byproducts. The use of immobilized photocatalysts with strong light absorption, chemical stability, and high surface area could greatly reduce the need for oxidant feedstocks and improve the affordability of AOPs.

Here, we report on the fabrication and characterization of photocatalysts immobilized on high-surface area supports for UV-driven water treatment. Our supported photocatalyst is mesoporous which, combined with the macroporous (Φ >0.95) support, yields a high surface area beneficial for mass transport. By using UV transparent supports, we maximize the photon management in our reactor.

We test the long-term stability and efficacy of our supported photocatalysts to degrade model contaminants under illumination from both UV-A LEDs and germicidal UV-C lamps. Our immobilized photocatalysts produce >100 μM *OH in 20 seconds in a flow reactor and achieve an Electrical Energy per Order (EE_O) of <0.6 for the degradation of Rhodamine B. We also examine the impact of water quality (pH, TOC, scavengers) and propose mechanisms for pollutant breakdown. Lastly, we will discuss the reactor design and relevant transport conditions for our supported photocatalysts, and the improvements necessary to scale our material for practical applications.

56e Wastewater Treatment Using Anaerobic Membrane Bioreactors with Electrolytic Regeneration (AMBER)

Maryam Amouamouha, Travis Walker

The constant release of different kinds of wastewater into the environment has significantly influenced the freshwater resources worldwide. Innovative recycling technologies using biological, chemical, and physical systems have been introduced with the aim of protecting humans, wildlife, and the environment. Our proposed system, called an anaerobic membrane bioreactor with electrolyte regeneration (AMBER), is a smart, stand-alone, self-sustained technology for wastewater treatment. The objectives of the AMBER development are to upgrade the cost and performance of the existing technologies and to meet the requirements of the standards in reusing and recycling water. AMBER is a bio-electrochemical process that is expected to be simultaneously efficient in water recovery, biogas augmentation, and fouling mitigation. In this work, we identified the wastewater-treatment performance and appraised the filtration behavior of AMBER. Considering the unique properties of AMBER hybrid system, circular economy goals, and ensuring sustainable wastewater management according to the 6th goal of sustainable development concept, in this investigation we have focused on this technology to achieve a novel package system. The objectives were to enhance the stability and efficiency of the anaerobic membrane bioreactor. The integration of the electrolytic process with the ABR (EABR) using a pair of iron electrodes enhanced the removal of contaminants in the ABR while successfully maintained pH in the optimum range for anaerobic digestion (6:8 to 7:2). The results showed that the AMBER configuration considerably enhanced permeate flux, as it reduced the deposition of extracellular polymeric substances (EPS) on the surface of the nanocomposite membrane, leading to a reduction in membrane fouling.

56f Green Oxidation Processes for Industrial Wastewater Treatments

Kun-Lin Yang

In industrial wastewater treatment processes, Fenton's reaction is an effective method for the treatment of dye wastewater, phenols and other recalcitrant pollutants which cannot be removed by using biological processes. In the reaction, H_2O_2 is injected into the wastewater in the presence of ferrous ions to produce powerful hydroxyl free radicals (·OH). However, Fenton's reaction only happens in a very narrow pH range (2.8-3.2). At neutral pH, ferrous ions precipitate and become sludge. Alternatively, H_2O_2 can be combined with UV and ozone to produce hydroxyl free radicals, but the treatment cost is very high.

To address the issues in Fenton's reaction, we developed new metal-based catalysts which can be used in industrial wastewater treatment by using H_2O_2 as an oxidant. The catalysts are able to complete a redox cycle in the presence of H_2O_2 and generate both free radicals and high-valence metal ions. These highly reactive species are able to convert hydrocarbons, especially unsaturated organic molecules, into organic acids and carbon dioxide with high molecular efficiency. These metal-based catalysts work over a wide range of pH (3-11) and do not produce any sludge. They are soluble and can be mixed with wastewater homogenously at parts-per-million (ppm) level at neutral pH to degrade the pollutants rapidly. Because of the fast reaction rate, the hydraulic residence time of the process in a wastewater treatment plant is usually less than one hour. They can also be immobilized on solid carriers to treat the wastewater continuously in a fluidized bed. Currently, the new process has been tested in a full-scale plant in Singapore and that leads to better treatment efficiency and 50% cost-saving compared to current treatment processes.

In addition to industrial wastewater treatment, the catalysts can also be utilized in manufacturing processes such as decolorization of vegetable oils and bleaching fabrics. They are also potentially useful as chemical disinfectant for the inhibition of algae growth and deactivation of bacterial endospores. There are many opportunities for the green catalysts to improve our lives in the future.

313a Synthesis, Transport Properties, and Benefits of Multifunctional Membrane Platforms for the Separation of Contaminants from Aqueous Environments

Francisco Leniz, Hailey Craft, Lindell Ormsbee, Dibakar Bhattacharyya

Challenges related to environmental remediation of water sources require of more efficient and selective separation technologies, supported by proper understanding of the separation mechanisms. Polyfluoroalkyl substances (PFAS) and heavy metals are examples of organic and inorganic compounds that must be removed. Separation processes using traditional technologies have some benefits and challenges when separating these pollutants from the environment. Benefits are created by design, as for example good adsorption affinity, high exclusion rates, high conversion rates, low energetic requirements, among others. Challenges arise from the limitation of the applicability of these benefits to a large range of targeted pollutants, as well as from the competition of the active sites with naturally present competitors, the fouling of organic matter, etc. Ultimately, these challenges can be reflected on a decrease in the processes' removal efficiency.

This work introduces the synthesis of multifunctional membrane platforms consisting of pore-functionalized microporous support thin-film composites. The main objective is to take advantage of the benefits from each separation mechanism, and their respective functionalities, to help overcome challenges with environmental remediation technologies. These thin-film composites consist of microfiltration membranes (about 100 nm pores) that can be pore functionalized with diverse responsive, strong acid, or strong basic polyacrylamide hydrogels, with a thin film on top that has nanofiltration (NF) characteristics. Characterization of thermo-responsive poly-N-isopropylacrylamide (PNIPAm) nanofiltration membranes using scanning electron microscope and x-ray photoelectron spectroscopy demonstrate the formation of the thin layer on top of the open microporous structure. The porous support from these membranes adsorbs organics (178 mg PFOA adsorbed/m^2 membrane at an equilibrium concentration of 70 mg/L), and the simultaneous exclusion from the NF layer allows separations of PFOA and the smaller size heptafluorobutyric acid from solutions containing 70 μg/L of these compounds at a high-water flux of 100 L/m^2-h at 7 bar. The synthesis and separation properties of strong acidic and strong basic functionalized porous support NF membranes will be also discussed. This Research has been supported by the NIEHS-SRC Program.

313b Pfas Degradation By a Thin Film Gas-Liquid Non Thermal Plasma Reactor

Narasamma Nippatlapalli, Rachel Gallan, Radha Krishna Murthy Bulusu, Robert Wandell, Bruce R. Locke

Contamination of groundwater by per- and polyfluorinated substances (PFAS) is an emerging worldwide environmental problem and current water treatment methods for dealing with PFAS use the non-degradative methods of activated carbon and reverse osmosis for physical removal from the contaminated water. In the present work we utilize a non-thermal plasma chemical reactor to degrade the PFAS: 8-carbon (C8) (perfluorooctanoic acid – PFOA), 6-carbon (C6) (perfluoro hexanoic acid - PFHxA) and 10-carbon (C10) (perfluoro decanoic acid - PFDA). The non-thermal plasma, a type of ionized gas, is generated by a high voltage electrical discharge in a reactor with flowing gas (argon) and liquid (water). The plasma discharge channel propagates along the interface between the gas and the thin liquid water film where the generated reactive chemical species, including reductive and oxidative species, degrade the PFAS into smaller carbon chain daughter products and mineralization products such as carbon dioxide and fluoride (F-). We have quantified PFAS degradation, mineralization, and product formation for different feed concentrations (10 ppb to 1 ppm) in deionized water solutions and for water conductivity up to seawater. The effect of various parameters such as pulse frequency (kHz), liquid flow rate (mL/min), solution conductivity (mS/cm), and radical scavenger addition are investigated. The results showed that based on chain length of PFAS, a shift in optimal mineralization was observed. Longer chain length (C10) PFAS resulted in faster removal rates for than shorter chain length (C6). Through optimization of the plasma reactor operating conditions (pulse frequency and pulse width) we have achieved 90-75 percent degradation of these selected PFAS. Maximum defluorination of PFDA of 49.89%, and PFHxA of 64.76% were obtained at 1 mL/min of flow rate due to increased exposure time to plasma generated reductive and oxidizing species. PFDA (C10), and PFHxA (C6) removal efficacy was enhanced by improving the surface activity by adding NaCl to increase conductivity between 0.25 mS/cm to 40 mS/cm. At 3 mS/cm, PFHxA showed 55.81% defluorination while C10 showed 40.85% at 25 mS/cm. The effect of different liquid phase scavengers such as ethanol for hydroxyl radicals and chloroacetate for aqueous electrons are studied. Scavenging both radicals and aqueous electrons decreased the defluorination rate of PFHxA by 48.31% while PFAD was reduced by only 3.84%. Overall, the results demonstrate high efficiency of PFAS degradation and mineralization, and thus indicates that the thin film gas liquid non-thermal plasma is a promising method for PFAS degradation.

313c Pfas Surfactant Sequestration By Binding to Functional Polymers

Paschalis Alexandridis, Samhitha Kancharla, Aditya Choudhary, Dengpan Dong, Dmitry Bedrov, Marina Tsianou

Surface active per- and polyfluoroalkyl substances (PFAS) find niche applications because of their high chemical and thermal stability, their incompatibility with both water and hydrocarbons, and their unique ability to render surfaces non-stick. However, several widely used PFAS surfactants have been found extremely resistant to degradation, accumulate in the environment, and have long half-lives in humans, consequently causing great concern [https://doi.org/10.1016/j.jhazmat.2021.128137]. In the context of developing materials and processes for sequestering PFAS surfactants from aqueous media, we research how such surfactants associate with (bind to) polymers and surfaces. To this end, we utilize complementary experiments (small-angle neutron scattering, SANS, with contrast variation) and modeling (molecular dynamics, MD), and present here examples on how the notorious PFAS surfactant perfluorooctanoate (PFOA) associates with poly(ethylene oxide) (PEO) homopolymer and amphiphilic block copolymers [https://doi.org/10.1016/j.jcis.2021.10.176]) in aqueous solution, and with cyclodextrin-based polymer networks [https://doi.org/10.1021/acsmaterialslett.2c00168]. A detailed description emerges on how PFAS molecules organize around the different parts of the polymers, which is used to rationalize the capacity of these polymers for PFAS sequestration.

313d Pfas Fate and Transport Study in Contaminated Sediments Utilizing Capping Methods and Sorbent Amendments

Prashik Manwatkar, Hossein Dallalzadeh Atoufi, David Lampert

Perfluorooctanoic acid (PFOA) and perfluorooctane sulfonate (PFOS) are two important organic chemicals of the per- and polyfuoroalkyl substances (PFAS) group that have contaminated land, water, and the air since 1950. The continuous release of PFAS from the surface of land into water is not easy to forecast and an appropriate treatment method needs to be economically viable since there are currently around 42,000 suspect industrial and municipal sites in the United States. For a true reproduction of real-world pollution patterns, we constructed polypropylene tanks, performed laboratory-based experiments, and analyzed the samples using EPA method 537. In this study, we examined the fate and transportation of long- and short-chain PFAS, including PFOA, PFOS, and perfluorobutanesulfonic acid (PFBS), from sediments, adsorbent media, and sands under overlaying water tanks. Activated carbon (GAC), biochar (BC), and fluorosorb (FS) were also added between the contaminated sediments and the sand layer in order to observe capping effectiveness. As one of the best ways to treat contaminated sediments on a large scale, adsorbent beds may reduce contaminants migration and support the degradation of contaminants. We found that all three chemicals were able to pass through the adsorbent layers of 3-4 inches from 4-5 inches of contaminated sediments and reach the top surface of the beds (25-30 inches). In the top 5-7 inches, PFBS concentration varies from 0.28 ppb to 0.78 ppb for all adsorbent tanks for 7 days. Whereas the bottom contaminated sediments concentrations of PFBS were 8517.6 ppb to 9480.7 ppb. We also observed the concentrations at top ports increased by 0.59 ppb to 2.31 ppb in 21 days, and ultimately, 0.58 ppb to 7.07 ppb in 69 days. While PFOA and PFOS found different metabolites in all layers, they provided noticeably lesser concentrations in contaminated sediments compared to PFBS. Further, the results of this study can be useful for validating the contaminant transport model predictions by identifying linear or nonlinear sorption equilibrium processes and diffusion-dispersion processes in sediment, sand, and various adsorbent media.

313e Superfine Activated Carbon-Functionalized Adsorptive Thin-Film Nanocomposite Membranes for Enhanced Pfas Removal from Water

Medha Kasula, Milad Esfahani

Due to the widespread production and use of per- and polyfluoroalkyl substances (PFAS), as well as their ability to move and persist in the environment, surveys conducted by the Centers for Disease Control and Prevention (CDC) show significant human exposure to PFAS. Membrane filtration and adsorption process are the two main effective separation and removal methods for PFAS removal from water. Thin-film composite (TFC) nanofiltration (NF) membranes showed an acceptable (<90%) removal of PFAS. However, the presence of rejected PFAS in the concentrate stream needs another separation process. In this study, we report the fabrication of adsorptive thin-film nanocomposite (TFN) NF membranes functionalized with super fine (<100 nm) activated carbon (SFAC) for enhanced PFAS removal by adsorption during the water filtration. The NF membrane was fabricated on a poly (ether sulfone) (PES) support layer with a polyamide (PA) selective layer functionalized by immobilization of AC inside the PA via interfacial polymerization. The isotherm adsorption of SPAC showed 93% removal of Perfluorooctanesulfonic acid (PFOS) in DI water. The bare NF membrane showed 12 ($L.m^{-2}.h^{-1}.bar^{-1}$) water permeability, around 60% NaCl rejection, more than 99% salt $MgSO_4$ rejection, and 90% PFAS removal. Based on the recorded results, the SPAC TFN membranes show enhanced PFAS removal due to the hybrid mechanism of size exclusion and adsorption without sacrificing the membrane permeability and antifouling properties.

249a Optimization of a Wastewater Cotreatment Process for Blowdown and Produced Waters with Economic and Sustainability Analyses

Hunter Barber, Victor Alves, Fernando V. Lima

Water is cycled in thermoelectric power plants and purged as cooling tower blowdown to alleviate contaminant accumulation. Fresh water is then used to regenerate the amount lost by the purge as well as additional evaporative losses. In the fracking process, high salinity wastewater is extracted from oil and gas wells during the process. As an inexpensive option, the produced water is disposed of via deep well injection. In contrast to the capital driven energy processes they accompany, lower effort is devoted to innovative and sustainable design practices for industrial wastewater treatment. This work evaluates the potential of mixing blowdown and produced waters to improve the overall process design. The concept is founded on exploiting complementary chemistry of contaminants to reduce the softening chemical demand in the combined process. The proposed cotreatment process design incorporates additional goals in reducing energy demand, increasing power plant water recycling, and producing saleable byproducts.

A cotreatment process model has been established using a combination of water treatment and chemical process simulators.[1] The model utilizes rigorous thermodynamics to capture extensive reactions and equilibria occurring in the mixing and softening of the waters. This plays an integral role in maintaining high model fidelity but increases computational expense associated with process simulation. This developed model when employed for process optimization also results in a highly nonlinear problem that is limited by simulator capabilities, which require fixing the modular flowsheet structure. Simulation information is then exported to additional tools in CAPCOST[2] and GREENSCOPE[3] to evaluate the results using economic and sustainability correlations and metrics. These metrics can holistically evaluate the performance of the design but further burden the computational expense of the process framework. With variations in potential operating decisions and combinatoric options for the formulated process flowsheet, it becomes difficult to infer best design cases and mixing ratios of the wastewaters.

These challenges together pose a complex and computationally expensive optimization problem. This work presents optimization techniques and model reductions to accommodate sufficient model fidelity, optimization compatibility, and perspectives on multiple process objectives. A computationally efficient reduced process model is used to generate a large data set at varying conditions and designs. Using this set, a Kriging metamodel based on literature methods[4] is generated and trained as a surrogate for optimization to minimize the levelized cost of water. Optimized conditions from the surrogate model are applied to the comprehensive process framework to extensively determine economic and sustainability metrics. The results are used to ultimately discuss the tradeoffs and viability of a more sustainably-focused cotreatment design.

References

[1] Barber, H. and Lima, F. V. (2021). "Modeling, Simulation and Optimization of a Synergistically Mixed Blowdown Water and Produced Water Wastewater Treatment Process". *Oral presentation in 2021 AIChE Annual Meeting, Boston, MA.*

[2] Turton, R, Shaeiwitz. J. A., et al. (2018). CAPCOST: Analysis, Synthesis, and Design of Chemical Processes, 5th Edition. *Prentice Hall, New Jersey.*

[3] Ruiz-Mercado, G. J., Smith, R. L., andamp; Gonzalez, M. A. (2012). Sustainability indicators for chemical processes: I. Taxonomy. *Industrial andamp; Engineering Chemistry Research*, *51*(5), 2309-2328.

[4] Caballero, J. A., andamp; Grossmann, I. E. (2008). An algorithm for the use of surrogate models in modular flowsheet optimization. *AIChE Journal, 54(10)*:2633–2650, 2008.

249b Hydrothermal Gasification of Concentrated Blackwater to Produce Energy and Low-Tier Water

Sean Thompson, Nathaniel Michael, Andrew Wagner, Alex D. Paulsen

The production of energy and low-tier water from blackwater was demonstrated at bench-scale through solids concentration, hydrothermal gasification, and subsequent wastewater treatment. Current wastewater disposal in remote environments is complex, expensive, and requires significant energy and resources. Conventional gasification of wet wastes is not feasible due to the high energy input required to remove and process water. However, Mainstream has developed a novel utilization of wet waste using combined hydrothermal gasification and physical water treatment systems which provides utility and relieves the disposal burden. This process is based around concentrating the solids through high-efficiency hydrocyclones, enabling the majority of the blackwater solids to be concentrated into a small fraction of the water. Ensuing gasification experiments showed >99% of organic material was converted into a low-BTU, flammable gas. The remaining water was then treated to meet low-tier water criteria. Energy analysis revealed that the system can produce net energy at a wastewater processing rate of over 1,000 gal/day, with >95% of incoming wastewater treated to be suitable for low-tier applications. Based on the bench-scale testing and system analysis, a preliminary full-scale design was developed to integrate subsystems and estimate costs. This presentation will discuss the novel water treatment process, energy and mass balances, hydrothermal gasification parameters and yields, and overall economics.

249c Response Surface Optimization for the Development of a Compact, Rapid, and Zero Liquid Discharge Wastewater Treatment System

Ramalingam Subramaniam, Alicia Viera, Ian Ivey, Jess Fike

Developed countries most commonly employ activated sludge treatment plants or utilize aerobic digestion in another manner to reduce solid content in municipal wastewater. These types of centralized treatment facilities are not feasible to construct or maintain in many places, and demand for new water conservation solutions is also growing globally. Zero-liquid discharge wastewater treatment, in which direct flocculation is the primary treatment step for solid-liquid separation, could serve as a rapid, compact, and highly efficient system for treating municipal wastewater in locations where abundant freshwater and skilled operating personnel is not available.

In this work, the direct flocculation process was used with a cationic polymer to flocculate raw municipal wastewater and the effects of manipulating process conditions were examined. A 5 level, 4 factor central composite design of experiments was performed to analyze and model the response of TSS and COD reduction as a function of flocculant dose, mixing time, mixing speed, and polymer solution concentration by applying response surface methodology. Direct flocculation proved to be an efficient and rapid method for inducing the settling of solids in raw wastewater.

249d Advanced Wastewater Treatment and Management for Water Circularity, Reuse and Recycling

Tapas K. Das, Murilo Innocentini

Water resources cannot, as they have in the past, be forever expanded to accommodate ever-growing consumption. In the last 75 plus years, world population has tripled while per capita consumption has more than doubled, resulting in a 600% increase in total demand. For industrial, municipal, and agricultural water users alike, the growing strain on water resources has been raising costs for freshwater, wastewater treatment, pollution mitigation, and stormwater discharge. Up to now, the conventional approach has focused on complying with regulations and finding specific solutions to specific needs: water for industry, irrigation, human consumption, energy generation (food-energy-water nexus), and ecosystem preservation, without seeing each element as part of a larger system that needs to be managed holistically.

"Water" is the next oil" particularly in developing countries with population increasing! The increasing scarcity of water coupled with escalating cost of fresh water and its treatment has prompted industry and other users to consider water conservation, reuse, and recycling. Incorporating advanced technologies such as membrane bioreactors (MBRs), reverse osmosis, microfiltration, nanofiltration, and so on, would result in reclamation and reuse of water and less environmental damage, but to what degree, and with what trade-offs and the cost per million gallons? We use water to clean everything but how do we clean our water to maintain high water quality standard? To answer these questions, this presentation includes few mini case studies on industrial and municipal wastewater treatment systems with a focus on short-term and/or long-term effects and/or benefits of reusing and recycling waters on economics (life cycle cost, circular economy), energy, public health, environment, and water-ecology.

A few streamlined life cycle assessments (LCA) will be discussed on water reuse/recycling practices in municipality wastewater treatment plants, pulp and paper, food processing, pharmaceuticals, power generating industries, internal water reuse and recycling within industrial parks (wastewater minimization toward zero liquid discharge); combination of drinking-water and wastewater unit processes and treatments to produce usable water from domestic and municipal wastewater sources; high-rate or rapid filtration using granular media; slow sand filtration (appropriate technology for rural communities); ultraviolet vs chlorine disinfection for wastewater reuse, agricultural, land applications and direct groundwater recharge of reclaimed water, and water augmentation to ravines and rivers during summer low flow.

Key References:

Das, T.K. (2003). "Introduction to Life Cycle Assessment as Applied to Water Reclamation and Reuse", Presented at the Conference - Ecological Engineering for Integrated Water Management: Designing Urban and Industrial Watersheds, Harvard University School of Design, Cambridge, Massachusetts, Oct 30 – Nov 2.

Designing Urban andamp; Industrial Watersheds (ecological-engineering.org)

Das, T.K. (2005). *Toward Zero Discharge: Innovative Methodology and Technologies for Process Pollution Prevention,* John Wiley and Sons, Hoboken, NJ.

Mihelcic, J.R. and Zimmerman, J.B. (2014). *Environmental Engineering: Fundamentals, Sustainability, Design*, 2 ed., John Wiley and Sons. Hoboken, NJ.

Das, T.K. (2020). *Industrial Environmental Management: Engineering, Science and Policy*, 1st ed., John Wiley and Sons, Hoboken, NJ.

249e Treating Hydraulic Fracturing Produced Water By Electrocoagulation

Mahmood Jebur, Yuhe Cao, Mahdi Malmali, Xianghong Qian, Ranil Wickramasinghe

Hydraulic fracturing oil and gas produced water is frequently highly impaired. While it is often deep well injected, there is a great deal of interest in treating and recovering this water for beneficial uses. However multiple unit operations are needed if these wastewaters are to be recovered. Electrocoagulation is considered a promising pretreatment technology. Herein, we have investigated the use of aluminium electrodes for electrocoagulation as a pretreatment operation. The effects of electrode arrangement, applied current, reaction time, pH, and inter electrode distance on the quality of the treated water have been investigated.

The results obtained here indicate that electrocoagulation can obtain good removal of turbidity (95%), total suspended solids, TSS (90%), and total organic carbon, TOC (69%) by carefully choosing the reaction conditions. Sedimentation was used to separate the treated water from the sludge. The performance of the electrocoagulation process depends strongly on the quality of the feed water. The viability of a practical continuous electrocoagulation process will depend on the volume (footprint) of the reactor which in turn will depend on operating conditions and the quality of the feed water.

249f Impacts of Invasive Species Resulting from Canals

Robert Peters, Tapas Das

A canal is a human-made waterway allowing ships to pass from one body of water to another. Examples of major canals include the Panama Canal, the Erie Canal, and the Suez Canal. Canals are also used to transport water for irrigation and other human uses. The Panama Canal connects the Atlantic Ocean and the Pacific Ocean. Construction of the Panama Canal began in 1904; construction was completed in 1914. The Suez Canal connects the Mediterranean Sea and the Red Sea; construction of the canal began in 1859 and was completed in 1869. Although canals greatly reduce the time of transport of commodities, it also has resulted in invasive species being transported from one body of water to another. This is an environmental problem that has implications on the healthy development of the ecological biosphere. Canals serve as a conduit for animals and plants to interact in new areas. Species can be picked up in their native area and brought to a new location on the other side of the world.

This presentation summarizes the results of a comprehensive technical literature review, addressing the impacts of invasive species on various waterways connected through major canals. Examples of invasive species investigated include: puffer fish (*Torquigener flavimaculosus*), mollusks, green crab (*Carcinus maenas*), Asian bighead carp (*Hypophthalmichthys nolibis*), silver carp (Hypophthalmichthys molitrix), sea lampreys (*Petromyzon marinus*), silver-cheeked toad-fish (*Lagocephalus sceleratus*), Rabbit fish, Erythrean mytilid mussels, spiny waterflea (*Bythotrepes longimanus*), fishhook waterflea (*Cerapagis pengoi*), mosquitos (*Culux* and *Aedes* species), and aquatic plants such as European watermilfoil (*Lythrum salicaria*), water chestnut (*Trapa natans*), tall invasive grass (*Saccharum sponteneum*), Western Atlantic caprellid amphipod (*Paracaprella pusilla*), geen alga (*Caulerpa prolifera Lamouroux*), and perennial herbaceous species (*Saccharum spontaneum*).

495a Techno-Economic and Environmental Analysis of the Conversion of Food Waste to Renewable Energy

Yunzhi Chen, Elizabeth Grace Pinegar, Kody Powell

Over 103 million tons of food waste are generated in the United States each year, making up over 24 percent of municipal solid waste (MSW) sent to landfills [1]. However, landfill costs are rising due to a lack of suitable landfill sites; furthermore, food waste often has the highest methane potential among MSW elements, accounting for a large portion of fugitive greenhouse gas (GHG) emissions from landfills [2]. Anaerobic digestion (AD) has the potential to be a more sustainable and value-added option for food waste treatment [3].

This study focuses on the techno-economic and environmental assessment of AD as a food waste solution. A city with a million population is assumed, and a percentage of food waste that would otherwise go to landfill may be collected by a stand-alone anaerobic digestion facility. As shown in **Figure 1**, four typical pathways are evaluated: flare, pipeline natural gas (PNG), onsite power generation including combined heat and power (CHP), and compressed natural gas (CNG). The current technology of different modules is discussed in this study. With reasonable assumptions, economic indicators such as simple payback period (PBP) and net present value (NPV) are calculated, and sensitivity analysis is performed.

The study finds that tipping fees play a significant economic role for the food waste treatment plant, with a general tipping price of more than $30 per ton required for a positive NPV. With conventional AD technology, the cost of biosolid treatment accounts for a substantial portion of the AD operation costs, but advanced AD technology such as thermal pretreatment (THP), solar drying, and thermal drying could potentially improve biosolid quality and lower disposal costs. CNG offers the greatest economic performance of the four pathways, yet market demand may constrain its use. With certain policy incentives on renewable natural gas (RNG) and electricity prices, PNG and onsite generation paths have a positive NPV, and advanced biosolid solutions potentially enable a positive NPV without policy support in the future. Food waste AD plants release significantly less GHG than landfills, according to environmental analysis, implying that they can play an important role in the environmental performance. This study suggests that AD will be a feasible and more sustainable food waste solution than landfill from an economic and environmental perspective.

References

[1] United States Environmental Protection Agency, "2018 Wasted Food Report," no. November, p. 42, 2020, [Online]. Available: https://www.epa.gov/sites/production/files/2020-11/documents/2018_wasted_food_report.pdf.

[2] A. Badgett and A. Milbrandt, "Food waste disposal and utilization in the United States: A spatial cost benefit analysis," *J. Clean. Prod.*, vol. 314, no. August 2020, p. 128057, 2021, doi: 10.1016/j.jclepro.2021.128057.

[3] S. K. Pramanik, F. B. Suja, S. M. Zain, and B. K. Pramanik, "The anaerobic digestion process of biogas production from food waste: Prospects and constraints," *Bioresour. Technol. Reports*, vol. 8, p. 100310, Dec. 2019, doi: 10.1016/J.BITEB.2019.100310.

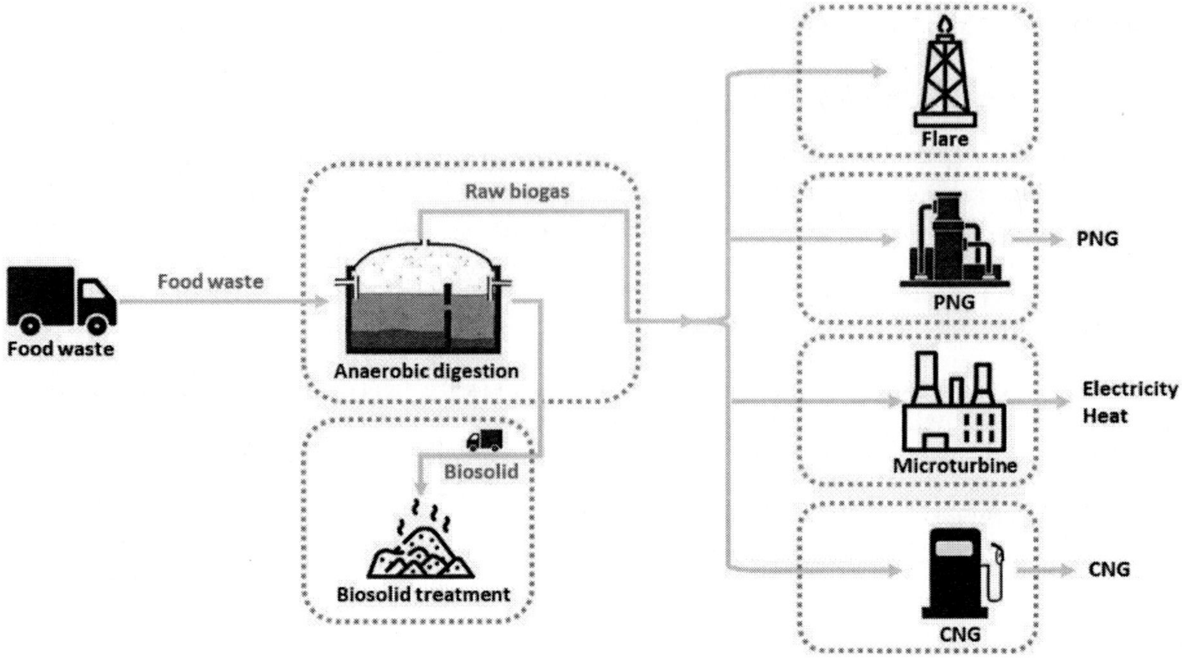

Food waste
Food waste
Raw biogas
Anaerobic digestion
Biosolid
Biosolid treatment
Flare
PNG
PNG
Microturbine
Electricity
Heat
CNG
CNG

495b Resource Recovery from Solid Waste Anaerobic Digester - Performance, Energy Consumption and Cost

Yupo J. Lin, Deepak Dugar, Christopher W. Simmons

Organic waste streams, including agricultural residues and food waste, are predicted to significantly increase in the coming decades due to rapid urbanization. One promising technology to reduce and reuse organic waste is anaerobic digestion (AD). To enable wide and economical deployment at small scale AD, higher value products from recovered waste organic acids are necessary compared to conventional, limited value biogas product.

In this presentation, we will discuss the engineering design, installation, and field operation of organic and inorganic salts recovery technologies from the liquid phase of solid waste anaerobic digester. The resources recovery simultaneously prevented organic waste from reaching landfills and potentially produced a sustainable green alternative to petrochemicals at a cost-competitive price. Process performance index metrices of the recovery will be discussed. These metrices were used to optimize the operation condition for maximum productivity, minimum energy consumption and capital equipment footprint. Technology viability of the recovery technology was assessed in a pilot-scale technology demonstration using the influence from a field AD. We will also discuss the robust operation in a real-world environment and the mitigation actions executed in several events of process upsets.

495c Predicting the Role of Reactive Nitrogen Intermediates during Hydrothermal Liquefaction of Food Waste

Heather LeClerc, Rasha Atwi, Amy M. McKenna, Michael T. Timko, Richard H. West, Andrew R Teixeira

Hydrothermal liquefaction (HTL) has emerged as a promising strategy for converting abundant, water-rich organic streams into an energy-dense, sustainable, biocrude. Despite major strides in improving oil yields and process intensification, a key underlying challenge persists in relating the composition of the feedstock to the ultimate fate of nitrogen, which is especially important for biocrude quality.

To elucidate how nitrogen partitions into the respective gas, aqueous, oil and solid hydrothermal products, food waste was treated under HTL conditions (15 wt% solids, 300 °C reaction temperature, and 60 minute reaction time) and elemental, chromatographic and mass spectroscopy analyses were performed on the products to establish dominant product classes and to close mass, carbon, and nitrogen balances. Liquid products (aqueous and biocrude) were found to contain nearly 60% of the nitrogen, with a majority of the biocrude-phase nitrogen in the form of amides. Ab initio thermodynamic and kinetic simulations on model reactants were used to evaluate potential reaction pathways involving reactive nitrogen intermediates.

A subsequent kinetic model was evaluated at reaction temperatures ranging 300 – 1000 K and for a range of feedstock compositions to identify primary reaction pathways. The Maillard reaction is revealed to be the dominant pathway for converting reactive nitrogen intermediates into observed products, especially those in which nitrogen is present as a heterocyclic aromatic. Subsequent product analysis by Fourier Transform Ion Cyclotron Resonance Mass Spectrometry (FT-ICR MS) confirms strong agreement among the experimental data, kinetic model and proposed Maillard pathways shown in **Figure 1**.

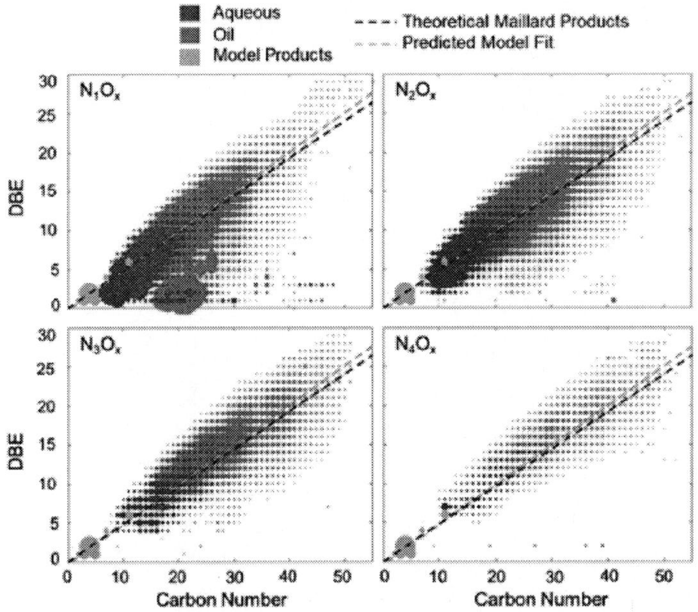

Figure 1. Double bond equivalency (DBE) versus carbon number for four nitrogen-containing heteroatom groups in the biocrude (red) and aqueous (blue) phases from food waste HTL at 300 °C for 60 minutes. The green dots represent the products from the kinetic model run with 18% glycine, 22% diene, and 59% aldehyde split evenly between the four starting aldehydes. Model feedstock concentration was chosen to match that of food waste HTL. Experimental data points are sized by relative abundance and model points by relative predicted concentration.

Processes to Enable Full Material Recovery from Municipal Solid Waste

Tommy John, Tommy John Engineering, Bandera, TX

Abstract

Recycling, or material recovery, is the key to sustainability. Depositing material in a landfill or combusting in a Waste to Energy (WTE) plant, as is a common practice for municipal solid waste (MSW), should be replaced with material recovery facilities (MRF). Sustainable material recovery from waste streams requires rigorous sorting and reliable, efficient processes for conversion of sorted streams into high value products. Plastic recycling is less developed than for other recyclables, but the Pyrolysis and Methanolysis processes extend the range of plastic that can be effectively recycled from waste streams. The remaining organic material following removal of all recyclables can be upcycled into valuable oxygenated chemicals.

A preliminary analysis indicates that a MRF system that can recover the highest value for recovered materials will produce revenue that is five to six times the revenue from typical landfill tipping fees, thus creating a sustainable revenue stream for the facility while maximizing environmental benefits.

Background

Sustainable waste management requires robust Material Recovery Facilities (MRF) to replace landfills and Waste to Energy (WTE) plants. In developing areas, MSW is often dumped directly into the environment (land or water), and should be replaced with material recovery. Landfills have historically been the lowest cost option for waste disposal. WTE is more expensive, but is often used in regions with a high population density and limited land. Both practices are difficult to permit and face considerable public opposition. Processing the full MSW stream in a MRF will maximize recycling rates and improve public cooperation with reducing waste, including plastic, in the environment. Gaining acceptance for anti-litter campaigns is easier than for recycling, which is more complicated.

In addition to the obvious loss of valuable materials, landfills have other disadvantages. Landfills require a lot of land, and are often in a less accessible location requiring transportation of waste for a considerable distance. Although they can be constructed and operated to minimize environmental impact, the capture of landfill gas is limited to 70 to 80% by the need to limit air ingress to prevent landfill fires, resulting in unavoidable but substantial emissions of carbon dioxide, methane, and VOC. The landfill gas can be converted to natural gas or electricity, but the efficiency is low and carbon dioxide emissions are not abated.

WTE plants generate electricity using a steam cycle, but are limited by corrosion and fouling constraints to lower steam temperature and pressure, which, along with parasitic losses from emission control equipment, limits the net conversion efficiency to 20 to 30%. This is quite low compared to current combined cycle units that have an efficiency approaching 60%. Although WTE produces a high value product, the recovery efficiency is low, and capital investment is high.

Digital technology enables improved sensors, controls, monitoring, artificial intelligence (AI), and mechanisms, including robotics, thus continuing to rapidly improve the sorting technology used in MRF. Rigorous sorting can direct components of the waste stream for recovery as high value products, so it is paramount to have processes that effectively recover the high value products from the sorted materials. As with all energy and disposal facilities, a MRF based waste processing facility will require a large initial capital investment. In order to pay off the bonds or other financing of the investment, a high value for the recycled material must be collected in addition to a tipping fee. The value of the recycled material also aligns with environmental quality because, if a product has a high value, it is a scarce resource and/or it requires a lot of energy to produce.

The higher value recyclable materials include metals, paper, and plastic; the latter for which the technology is least developed. Waste plastics represent a unique challenge because of the diverse composition, but an opportunity because of the high value of the sorted components. No single process will optimally recover plastic wastes, but multiple processes, including mechanical, chemolysis, and pyrolysis can be combined to optimize recovery.

Transfer Stations

Transfer Stations are currently used to optimize logistics for moving collected MSW to disposal sites. In urban areas they are used to reduce haul distance for collection trucks, since, as noted, landfills are typically located outside the urban center. Rural counties and small town use them for collection for transfer to distant disposal sites, because a well-designed landfill is too expensive to serve a small population. More sophisticated MRF may require a larger scale for economy and location away from metro area justifying additional Transfer Stations integrated with MRF with rail or intermodal access.

Material Recovery Facilities

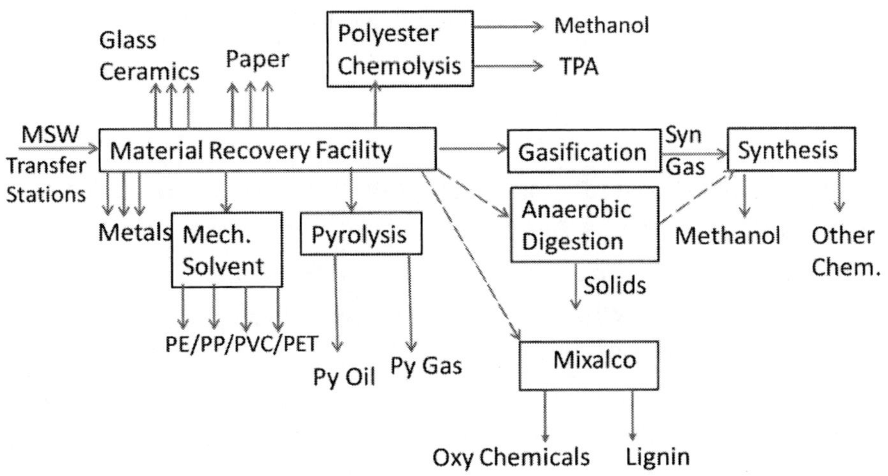

State of the art sorting can capture materials from the whole waste stream for processing for the highest value, including multiple streams for metals, paper, and plastics. Dense solid materials such as glass, ceramics, and rocks have minimal commercial value, but are removed to improve downstream processing. Metals, paper, and glass recycling are more advanced than plastic. The processing of plastic is challenging because of the wide range of products. Optimal recovery of plastic requires multiple processes because of the range of materials. Mechanical recycling is currently the most widely used, but requires high purity, clean material. Chemolysis can convert Polyesters, including PET into plastic feedstocks. Polystyrene can be converted to styrene by pyrolysis. Pyrolysis of mixed hydrocarbon plastics is particularly useful because it effectively converts many of the polymers remaining, after diversion into other processes, into Py oil, Py gas, and coke/asphalt, and tolerates contamination and mixtures. Processes are under development for other plastics.

The remaining materials, which may be more than half of the original waste stream, are mostly biomass derived, or similar to biomass in composition: food waste, yard waste, wood, and uncaptured paper and plastic. The stream will have a high moisture content, which is not necessarily a disadvantage for the downstream processes; gasification, anaerobic digestion, or carboxylate salts, but the product yield will be based on the dry material content.

It should not be inferred that all of the elements in the schematic above are co-located with the MRF. The exceptions are the plastic pyrolysis and remaining material recovery process because they are required to capture material that is not otherwise collected for recycle.

Waste Plastic Processing

Mechanical recycling, by which the material is ground to a powder and then extruded into pellets, is used to recover PE, PP, PET, PVC, and Polystyrene, but is limited by the requirement of high purity, clean material.

Pyrolysis Process

Pyrolysis of plastic waste will be an important part of the full material recovery process. This is not a new process; there has been a substantial amount of research, pilot plants, and even a few commercial plants implemented over several decades. For example, three commercial plants operated successfully in Japan in the 1990's, but lacked scale to be economic during a period of low oil prices and were closed down. Several plants are now in operation in Europe and North America, although still at a relatively small scale. It is currently receiving a lot of attention from major hydrocarbon processing operators and licensors. Although the process does not have a long operating history, it is less challenging than many of the successful processes in the petrochemical industry.

Pyrolysis processes extend the range of plastic that can be effectively recycled from waste streams (1). It performs best for hydrocarbon-based plastics, but it is effective with streams that have more diverse polymer compositions and contamination than can be tolerated with mechanical recycling. The main product is an oil stream with a composition that ranges from heavy oil to light naphtha that can be used as a refinery feedstock that is equivalent to petroleum. It can also be used as liquid feed to an olefins plant. An important secondary product is a gas stream that is rich in light olefins. It can be injected into an olefin plant for complete recovery, if the location has pipeline access to such a facility. Other-wise the gas can be compressed and the C3+ components condensed for storage and shipment in pressure tanks.

Polystyrene that cannot be mechanically recycled can be converted into styrene monomer by pyrolysis when fed separately from other polymers.

Chemolysis

Eastman Chemicals is successfully operating a new process (Methanolysis) to convert polyesters (including PET) into monoethylene glycol (MEG) and dimethyl terephthalate (DMT), intermediates for production of new polymers. With this process and the Pyrolysis process, almost the entire spectrum of plastic materials can be recycled or upcycled.

Processing of Remaining Organic Material

There are multiple options for processing the material remaining after the metals, glass, and almost all of the paper and plastic have been removed. It can be fed to an oxygen blown, high pressure gasification unit for conversion to syngas composed primarily of carbon monoxide and hydrogen that can be used to produce organic oxygenates. Methanol is an example, but it is a high-volume chemical produced with the economic advantage of economy of scale. Downstream oxygenated chemicals would have a higher inherent value because of additional processing and lower volume (less economy of scale for competitors).

There are other options for conversion of the unsorted materials, including anaerobic digestion, and fermentation into carboxylate salts (commonly known as the Mixalco process). The latter two (bio) processes will produce a residue of primarily lignin that would require further processing.

Value of Recycled Plastic Products

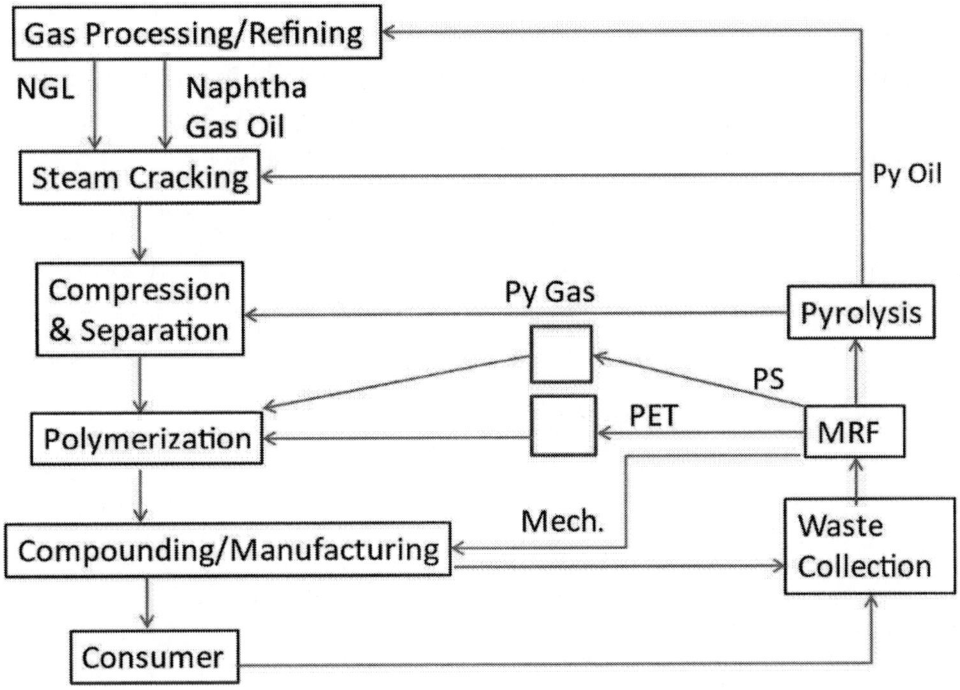

Polymer Cycle

The market value of the products produced by plastic recycling will depend on the uses. The chart above is a simplified illustration of the polymer supply chain for demonstrating the relative value and circularity of material from recycling processes. Pyrolysis complements rather than replaces Mechanical Recycling; the latter replaces polymer while the former is injected further upstream in the cycle. In the special case of polystyrene separated from the waste stream, it can be converted by pyrolysis into styrene monomer and used as feed to a polymerization process. Note that PET can similarly be recycled using chemolysis. The output of the Plastic Pyrolysis plant is py gas, py oil, and a limited amount of heavy residue (coke or asphalt, not to be confused with the heavy oil fraction).

In the special case of proximity to an olefins plant, the Py Gas can be fed to the compression section for full recovery of the stream. All of the constituents (alkanes, olefins, and hydrogen) are also in the cracked gas and will be recovered as product, feed, or fuel. Note that the pyrolysis plant does not have to be located within the olefins unit, but simply close enough for the gas to be transferred by a single pipe. The value of the py gas is the value of the individual constituents minus the cost of compression and processing. In the more common case of a remotely located Pyrolysis plant, the py gas can be compressed and cooled to partially condense a C3+ hydrocarbon stream containing primarily light olefins and alkanes that can be shipped in pressured vessels using ground transportation. In this case, ethylene, which is about 10% by mass of the py gas stream, would have only fuel value.

The use of Py Oil as a steam cracker liquid feed has been demonstrated. An alternative for any fraction that does not meet the specifications for steam cracker feed can be sent to a refinery. Compared to conventional crude oil, it is lighter, has more material in the range of transportation fuels, is low sulfur, and has lesser contaminants. Refineries have more options for processing and blending to meet fuel and feedstock specifications.

The following chart shows the market value of key commodities produced in the MRF system.

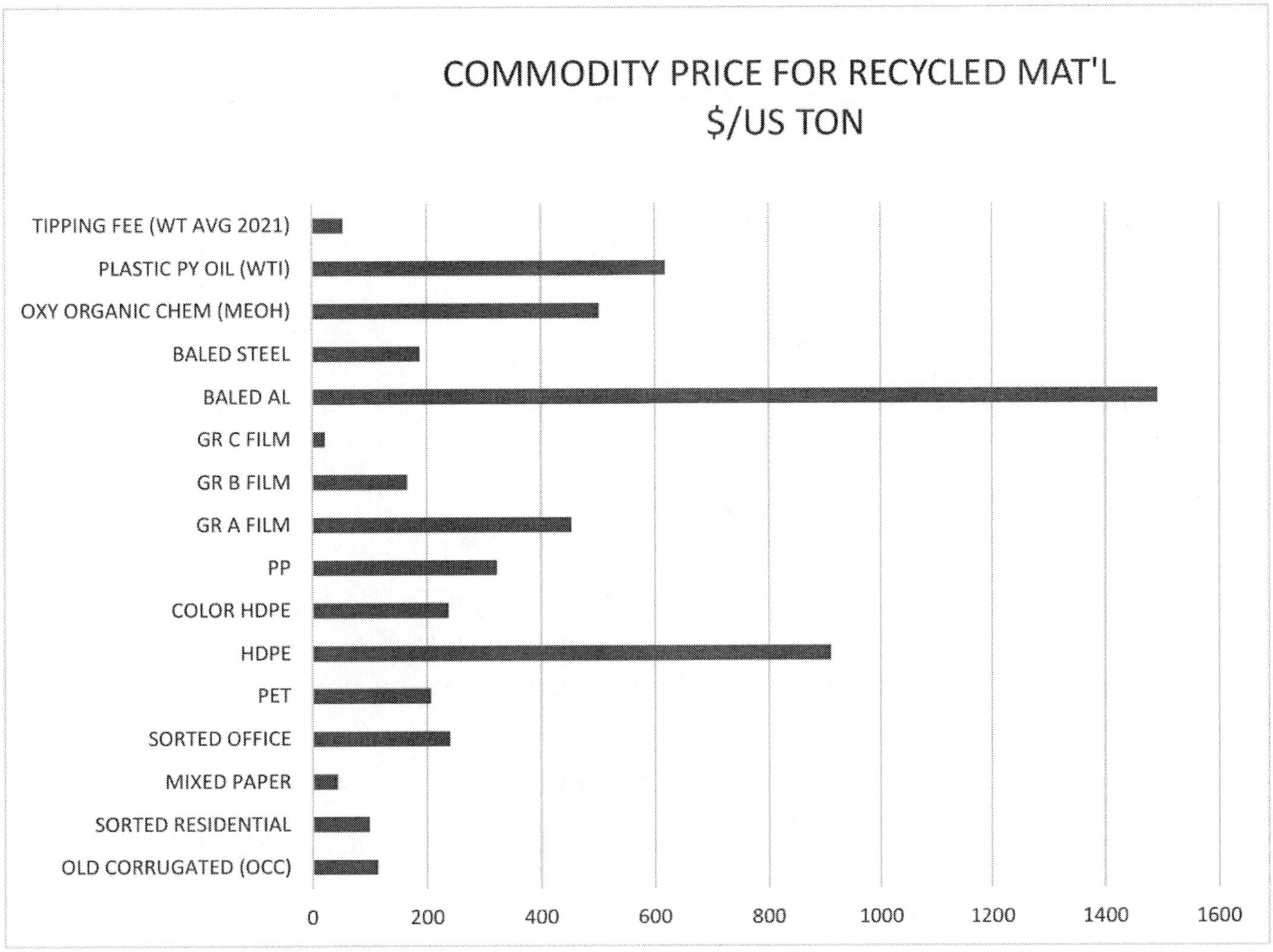

The tipping fee (54 $/ton) is the US Landfill average for 2021, a figure that has been relatively consistent for the preceding few years. The oil price is based on a recent value of $73 per barrel for West Texas Intermediate. The benchmark crude oil is slightly heavier and contains more sulfur than typical Py Oil, but the Py Oil is more olefinic. The oxygenated organic chemical price is based on recent US Gulf Coast contract pricing of Methanol. All other values are recently published prices for recycled commodities. Of course, all of these commodities will fluctuate in price, but the values are representative of pricing in the third quarter of 2022.

Estimated Revenue from Full MSW Recovery

The following chart shows the economic contribution of each element of the MRF system based on a representative composition of the municipal waste stream, an assumed recovery percentage for each process, and the commodity prices shown in the preceding chart.

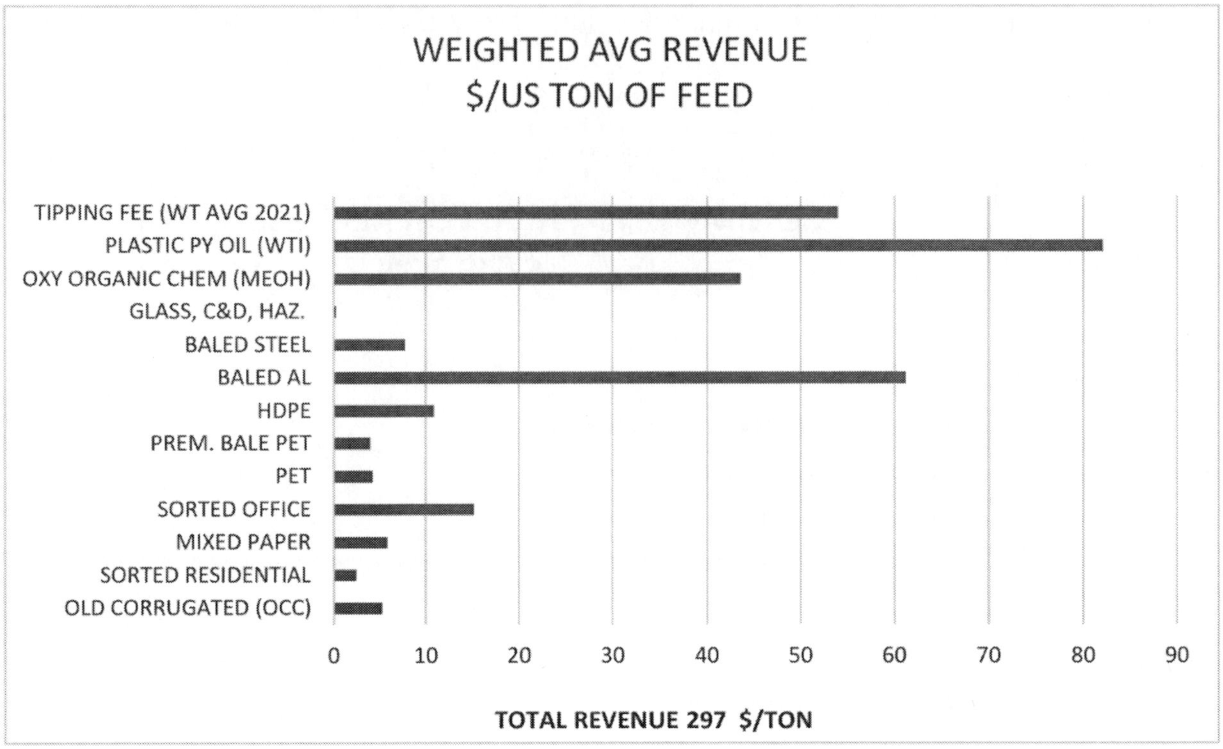

This indicates that the total revenue for the MRF is 5 to 6 times the current cost of disposal. Please note that the intent of this analysis is a preliminary, high-level view of the potential for full recovery of the MSW stream. While the figures may not be precise, the importance of capturing the highest value for each output is illustrated.

Conclusions

- Rigorous sorting and recovery of high value materials from the full MSW stream creates sustainable waste management.
- The environmental and economic benefits of material recovery are aligned.
- Waste plastics are valuable components of the waste stream, but lag other recyclables in recovery applications
- No single technology will optimize plastic recovery.
- Pyrolysis and Methanolysis are viable technologies for increasing the amount of plastic waste that can be effectively recycled.
- Multiple options exist for converting Organic Material to Oxygenated Chemicals.
- Material Recovery from the full waste stream is the best way to reduce waste, including plastic, in the environment.

References

1. John, Tommy, "Pyrolysis: Filling the Gap in Recycling Waste Plastic", Paper 639375, AIChE Spring National Meeting, April, 2022.

495f Decentralized Anaerobic Digestion Waste-to-Energy System for Food Waste Treatment and Fertilizer Production at East Coast Lagoon Food Village (ECLFV)

Yong Wei Tiong, Pooja Sharma, Yen Wah Tong

Singapore generated more than 660 million kg of food waste and the recycling rate has never exceeded 20% (National Environment Agency, 2020). As part of the valuable solution to manage food waste via environmental-friendly pathways, the development of sustainable solutions to treat food waste is desirable. One of which is the development of an on-site self-sustaining anaerobic digester systems that recycles food waste. Anaerobic digestion (AD) has been recognized as cost-efficient and environmentally friendly biotechnologies solution for renewable energy production, in the form of heat and power, organic waste management, and recovery of valuable nutrient rich digestate. Hence, this study is a pilot trial to test the feasibility of using on-site AD system to treat food waste generated at East Coast Lagoon Food Village (ECLFV), which is one of the famous hawker centre in Singapore. The electricity generated by the system is supplied back to support the ECLFV electricity needs, while the digestate is used as biofertilizer for cultivating plants at East Coast Park.

A multi-disciplinary approach is studied to realize the full potential of the environmental and economic benefits. In this study, three major tasks are undertaken simultaneously, with the aim to address the issue of food waste. Firstly, sorting and treating food waste using a customized on-site AD waste-to-energy system for food waste management and conversion of digestate to fertilizers. The performance of psychrophilic (27-29°C) versus mesophilic (35-37°C) AD are compared in a waste-to-energy system in terms of energy performance, treatment capacity, sludge reduction rate, and microbiological parameters. The electricity consumption and generation efficiencies were evaluated through the analysis of the whole system. Subsequently, the use of digestate for the cultivation of plants is further investigated in comparison with chemical fertilizers. Next, educating and influencing behavioural change on the separation of food waste by the public. As such, the social and psychological studies could encourage the public to participate in food waste recycling. Lastly, conducting life cycle assessment and cost benefit analysis to determine the environmental impacts and economic viability, as well as the sustainability of the technology.

The successful implementation of this on-site AD system results in ECLFV being a model for the rest of Singapore to learn from and emulate, with the target to achieve the vision of a Zero Waste Nation.

References

National Environment Agency. (2020). Love your food, waste less, save more. National Environment Agency, Singapore, Retrieved from

https://www.nea.gov.sg/our-services/waste-management/3r-programmes-and-resources/food-waste-management#:~:text=Food%20waste%20accounts%20for%20about,the%20744%2C000%20tonnes%20in%202019.

on March 9, 2022.

629a Utilization of Process Liquid from Hydrothermal Carbonization to Enhance Dehalogenation of Waste PVC

Vahab Ghalandari, Toufiq Reza

The United States generates more than 42 million metric tons of waste plastic waste every year. One of the main portions of waste plastic waste is waste polyvinyl chloride (PVC), which is a halogenated plastic. The rising awareness of the potential difficulties created by waste PVC plastics necessitates the development of a reliable method for treating them in a safe and ecologically friendly way, without the release of chlorinated organic compounds. Dehalogenation of waste PVC by hydrothermal treatment is a potential technology for creating a non-toxic and nearly completely closed system for waste PVC treatment. At subcritical conditions, water reacts with PVC, releasing chloride from the solid. According to several research, acids are often used to accelerate hydrothermal processes. In this work, the acidic process liquid of hydrothermal carbonization (HTC) of orange peel was employed to enhance the dehalogenation of PVC during hydrothermal liquefaction. The HTC was performed at 260 °C, and 1 h to produce the acidic process liquid with a pH of 4.2. On the other hand, HTL experiments were performed using a Hast-C Parr batch reactor at 300-350 °C, and 0.25-4 h. The results indicated that HTC process liquid improves HTL conversion with a maximum value of 3.2 wt%. The maximum HTL conversion was 50.1 %, which was achieved at 350 °C and 1 h. Moreover, HTL crude contains low chloride and significantly more carbon and hydrogen, resulting in a significantly greater heating value. Finally, the results indicated that the dehalogenation efficiency is high, which demonstrates HTL can significantly remove chlorine from PVC and move it into the aqueous phase. It means that there is an appropriate potential for upcycling PVC to clean solid fuel using HTL.

629c Gasification of Waste Plastic to Enable a Circular Economy

Ping Wang, Nicholas C. Means, Fan Shi, Jonathan Lekse, McMahan L. Gray

Plastic waste in the landfills/natural environment causes serious environmental problems and sustainability issue due to the loss of valuable and natural resources. In addition, disposal and transportation costs of the waste will increase as landfill space becomes scarce. The plastic waste is becoming a global concern and must be urgently addressed as plastic production is dramatically increased. Gasification of the waste plastic has recently gained attention mainly due to its flexibility. It generates syngas that can be used to produce energy, energy carriers (such as H_2) and chemicals (fuels, methanol etc.). In this study, the impact of plastic waste properties and operation conditions on gasification are investigated. The gasification of the plastic waste was performed in a drop tube reactor-QMS (quadrupole mass spectrometer)-cold tar trap system. Product gas yields, syngas conversion and syngas composition were evaluated for various plastic species at different temperatures, reaction times, and plastic/H_2O feed ratios. Experimental results showed significant differences between gasification products from various plastics. H_2 yield was elevated from gasification of low-density polyethylene and polypropylene. All these tests show a great potential of enhanced H_2/syngas production from waste plastic/landfill wastes.

629d A Generic Scenario Analysis of End-of-Life Plastic Management Concerning Chemical Additive Releases

John Chea, Austin Lehr, Kirti Yenkie, Joseph Stanzione III, Gerardo Ruiz-Mercado

Plastic demands and production have considerably increased the quantity of spent plastics, out of which over 90% is either landfilled or incinerated. Both methods for handling spent plastics are susceptible to releasing toxic substances that negatively affect air, water, soil, organisms, and public health. Therefore, improvements to the existing infrastructure for plastics management are needed to limit chemical additives release and exposure resulting from the end-of-life (EoL) stage of plastics. This work aims to develop generic EoL management scenarios to track and estimate the potential migration and releases of plastics additives throughout the plastics EoL stage and develop a software tool to evaluate the impacts of various plastics EoL processing techniques (mechanical recycling, incineration, and landfilling) on greenhouse gas emissions, energy footprint, and subsequent environmental releases. The current infrastructure and hypothetical scenarios involving enhanced recycling methods were analyzed. The major releases and emissions for every stage in the EoL phase were quantified, identifying considerable risks to workers and the environment. Subsequently, a python-based software tool was developed to provide insights into the impacts of various plastic recycling techniques. The potential hazards and risks identified in this research create an opportunity to design a safer closed-loop plastics recycling infrastructure to handle chemical additives strategically and support implementing sustainable materials management efforts to transform the U.S. plastics economy from linear to circular.

References

1. N. Hahladakis, C. A. Velis, R. Weber, E. Iacovidou and P. Purnell, An overview of chemical additives present in plastics: Migration, release, fate and environmental impact during their use, disposal and recycling, *Journal of Hazardous Materials*, 2018, **344**, 179–199.
2. R. Crompton, *Additive migration from plastics into food: a guide for analytical chemists*, Smithers Rapra Technology, Shawbury, New ed., 2007.
3. L. Law, N. Starr, T. R. Siegler, J. R. Jambeck, N. J. Mallos and G. H. Leonard, The United States' contribution of plastic waste to land and ocean, *Sci. Adv.*, , DOI:10.1126/sciadv.abd0288.
4. R. Jambeck, R. Geyer, C. Wilcox, T. R. Siegler, M. Perryman, A. Andrady, R. Narayan and K. L. Law, Plastic waste inputs from land into the ocean, *Science*, 2015, **347**, 768–771.
5. M. T. Brouwer, E. U. Thoden van Velzen, K. Ragaert and R. ten Klooster, Technical Limits in Circularity for Plastic Packages, *Sustainability*, 2020, **12**, 10021.

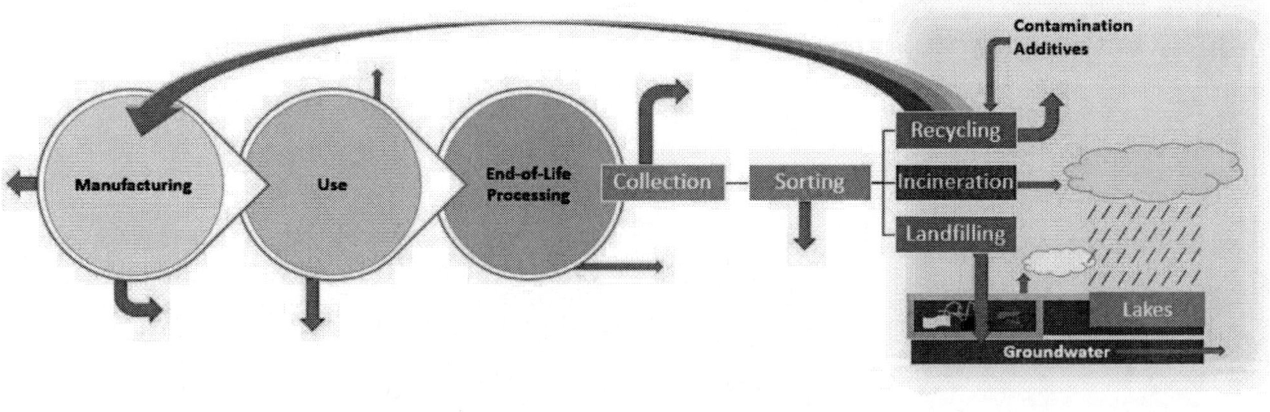

629e The Removal of Nanoplastics from Water with Lignin-Based Hydrophobic Deep Eutectic Solvents As Novel Extractants

Yuxuan Zhang, Jameson Hunter, Usman Abbas, Qi Qiao, Qing Shao, Jian Shi

As the emerging contaminations in aqueous systems, micro- and nano-plastics (MNPs) have been widely found to exist in the environment and even in organisms, which potentially threatens the ecosystems and human health. However, the studies evaluating the removal of these particles, especially the nano-sized particles, are limited. Herein, this work investigated the extraction of polystyrene and polyethylene terephthalate nanoparticles from water using lignin-based hydrophobic deep eutectic solvents (DESs). The removal efficiency was calculated via the pyrolysis-gas chromatography–massspectrometry (Py-GC/MS). The removal kinetics was also evaluated. Moreover, molecular dynamics (MD) simulations were applied to investigate the interaction among MNPs, water, and DES to reveal the removal mechanism. This work provides a possible solution to the recycling and reuse of MNPs as part of the grand challenge of plastic pollution.

629f When and Where Do Bio-Plastics Make Sense?

Rachael Rothman, Maryam Hoseini, Stuart Walker, Tony Ryan

Bioplastics are often touted as a solution to plastic waste, however when life cycle assessment is carried out this is often not the case and detailed analysis is needed to decide when and where bioplastics might make sense.

The term bioplastic can refer to plastics that are bio-based (i.e. made from a bio source) and/or biodegradable. A more helpful end-of-life term is compostable, but that can refer to both at home composting or industrial composting, which occurs at higher temperatures. Here we analyse pairs of plastics that can be used for the same purpose - PET andamp; bio PET, LDPE andamp; PBAT, polystyrene andamp; PLA - and compare their life cycle carbon footprints.

Producing plastic from a bio source requires a complex series of steps with associated energy input. As such, with current UK grid electricity and gas, the fossil based plastics assessed have lower carbon footprints than their bio alternatives. As grid electricity decarbonises over time, a break-even point can be reached and the bioplastic has a lower carbon footprint that the fossil plastic. We also analysis the impact of moving to hydrogen from natural gas as a source of heat.

It is not only carbon footprint that is important - the impact of the plastic if it becomes litter is also important. We will discuss metrics developed to measure the impact of littering.

The bio sources used to make bioplastics take in carbon dioxide during growth. When the bioplastic reaches end of life this carbon is most often released to the atmosphere, either through composting, incineration or landfill. If instead bioplastics were made to be durable, they would effectively lock in carbon for a long period of time and become a route for carbon sequestration. Over 1 gigatonne of carbon could be sequestered in plastics every year1.

[1] Ryan, A.J., Rothman, R.H. Engineering chemistry to meet COP26 targets. Nat Rev Chem 6, 1–3 (2022). https://doi.org/10.1038/s41570-021-00346-6

629g Many Happy Returns: Combining Environmental, Technological and Behavioural Sciences to Understand What Is Required to Make Reusable Packaging Sustainable.

Maryam Hoseini, Sarah Greenwood, Harriet Baird, Thomas Webb, Paul Mattinson, Patrick Fairclough, Rachael Rothman

There is currently over 7.3 billion tons of plastic in the Earth System, and it will grow to be 40 billion tons by 2050. Increased plastic consumption has resulted in high amounts of plastic waste ending up in the environment. Approximately 12 million tons of plastics accumulate in different environmental compartments (i.e. soil, freshwater and the ocean) every year. Plastic litter seems to be everywhere; it can be seen trapped along fences, scattered in the streets and along beaches and roadsides, where it pollutes water bodies, oceans, land and air. Plastic packaging can be particularly problematic as it accounts for 50% in weight of the total plastic waste in the world and it is often single-use and used "on the go", making littering and leakage from managed waste more likely. Increasing concern for sustainability compels citizens and enterprises to reduce waste and encourage recycling, reuse and remanufacturing of end-of-life products.

This presentation will discuss work from the "Many Happy Returns" project that examines the sustainable reuse of plastic packaging by identifying the best reuse model for different contexts, exploring the optimum materials, process and technologies for smart reusable plastic packaging systems and developing practices to encourage reuse. This paper follows on from our previous work which was awarded the IChemE Hutchison medal in 2022. Life cycle assessment, material and process technologies and willingness to engage analysis are used to evaluate the sustainability of the proposed reuse model. In particular, analysis of returnable Vytal bowls and cups as reuse containers for use in cafes as take-out packaging, will be presented. The analysed midpoint impact categories have been aggregated into the three main life cycle phases: production, use and End of Life (EoL). Then, they have been used to assess the environmental break-even point (BEP), i.e. the minimum number of uses necessary for a reusable bowls to be preferable than a single-use cardboard container. Reuse is highlighted as a key factor to increase sustainability, reducing carbon footprint and resource consumption. Results evidence all the conditions for reaching a BEP, allowing identification of possible strategies to improve the efficiency of reusable products alongside the willingness of people to obtain an environmental benefit.

629h Sustainable Use of Recycled Materials for the Cleaner Production of Value-Added Products

Appala Naidu Uttaravalli, Hrithika Ganta, Meghana Bodhimisetty, Gopi Katiki, Likitha Gadde, Sanjana Mutyapu, Bhanu Radhika Gidla

ABSTRACT

In the study, sustainable and low-cost products such as adhesive materials and composite products are prepared by using recycled materials such as post-consumer expanded polystyrene (EPS) and biomass material. To prepare the adhesive product, the EPS was dissolved in suitable solvents such as n-butyl acetate (n-BA), tetrahydrofuran (THF), methyl ethyl ketone (MEK), m-xylene and gasoline, and the corresponding adhesives are coded as A-B, A-T, A-M, A-X and A-G, respectively. Various physico-chemical properties such as solubility, viscosity, adhesive (shear) strength etc. of the in-house prepared adhesives are estimated. The cost-effective and eco-friendly composites are also prepared in the study using EPS and biomass (saw dust as a model compound), and the prepared composites are characterized as per the standard procedures. From the results, it is observed that MEK solvent offered maximum solubility of the EPS among the chosen solvents. The composite study revealed that post-consumer EPS and biomass materials have good candidate to the preparation of wood-plastic composites. We believe that the present study will be useful to the business visionaries / analysts to design a suitable process for mitigating the global problems associated with waste generation and its proper management.

Keywords: Recycled materials; Expanded polystyrene; Biomass; Adhesive; Composite

1. Introduction

In recent years, sustainable management has become the focus of research throughout the world. It includes, among many other things, the development of eco-friendly processes and products. The present research is undertaken on the same line, that is, to prepare eco-friendly and cost-effective products such as adhesive material and composites from the post-consumer (waste) expanded polystyrene (EPS) and biomass. The recycling of waste EPS will reduce the environmental pollution caused by its disposal and will lead to cleaner and greener earth; and also will offer value-added products for various possible applications.

From the thorough literature regarding the utilization of waste EPS, it is observed that a few studies are available on the development of an adhesive / glue material and wood-plastic composites from waste EPS and biomass for various applications. In recent years, the use of wood-plastic composites (WPC) has been lengthened into various applications which includes interior decoration, door and window making, flooring, automotive interior design etc. WPC materials can be prepared from various types of natural fibers and thermoplastics. To the best of our knowledge, from the literature, it is evident that the detailed studies pertaining to the preparations of wood-plastic composites from waste EPS using biomass is scares. Therefore, the objective of the present study is to prepare the sustainable materials out of recycled materials.

2. Preparation and characterization of adhesive product from recycled EPS

The adhesive product is prepared by dissolving the EPS in chosen solvents. Around 30 g of solvent is taken into a 100 mL conical flask and known quantity of EPS is added gradually into it under stirring condition. The addition of EPS is stopped when no more EPS dissolves in the solvent. The in-house prepared adhesives using EPS and different solvents were ascribed with different codes as mentioned in Table 1. The prepared adhesives are used to measure various properties, such as solubility, viscosity, adhesive strength etc. The solubility and viscosity (measured with Brookefield rheometer) data of the EPS adhesive are given in Table 1.

To find the shear/adhesive strength of the in-house prepared adhesives, shear strength test is carried out using a universal testing machine (UTM) equipped with 100 kN load cell. ASTM D5868 method is adopted to determine the adhesive strength of the adhesive products and wood material is used as substrate. To prepare the specimen for the shear test, the substrate material is cut into rectangular (101.6 mm length × 25.4 mm width) shaped structure and then two such structures were fixed in a single lap joint fashion with the prepared adhesives for a fixed overlap length. The specimen is kept under 5 kg load for a duration of 24 hrs to enhance the interaction between the adhesive and the substrate before performing the shear strength test. The test is performed by pulling the specimen at a speed of 2.0 mm/min under auto-trigger mode until the specimen breaks. In most cases, three numbers of specimen samples were prepared for the same experiment and the average value of the analysis results. The obtained results (shear strength values) of the adhesives are shown in Figure 1. The result shows that the shear strength of A-M adhesive is maximum among the four in-house developed adhesives. The order of shear strengths of the tested adhesives is A-X < A-B < A-T < A-M. The average value of the shear strength analysis results of A-X, A-B, A-T and A-M is 1081, 1372, 3791 and 4407 kPa, respectively.

3. Preparation and characterization of wood-plastic composites from recycled EPS and biomass

In the study, the EPS and sawdust materials are used to prepare the wood-plastic composites. The sawdust (size ≤500 μ) and EPS waste (size ≤1400 μ) were mixed at different ratios, and the EPS-sawdust mixture is used to make the composite. Compression moulding process is used to prepare the composites. The moulding temperature is maintained at 160 ±5 °C. The area of the prepared composite sheet is 225 cm^2 (15 cm × 15 cm). The prepared composite was used to estimate various physico-mechanical properties such as tensile strength, tensile modulus, %elongation, flexural strength, compression strength and water absorption etc. The images of prepared composites are shown in Figure 2. The effect of EPS loading on the properties of the in-house prepared wood-plastic composites are given in Table 2.

The data shows that the properties of the composites increased with the increase of EPS loading. Around 75% increase in tensile strength and %elongation is obtained with the increase in EPS loading from 40-50% by weight. It is also observed that around 60% improvement in tensile modulus; and around 50% improvement in flexural strength are obtained with the increase of EPS loading under present experimental conditions. It is also observed from the data that the water absorption values decreased with the increase of EPS loading.

4. Conclusions

In the study, sustainable and low-cost products such as adhesive materials and composite products are prepared by using recycled materials such as post-consumer expanded polystyrene (EPS) and biomass material. To prepare the adhesive product, the EPS was dissolved in suitable solvents such as n-butyl acetate (n-BA), tetrahydrofuran (THF), methyl ethyl ketone (MEK), m-xylene and gasoline, and the corresponding adhesives are coded as A-B, A-T, A-M, A-X and A-G, respectively. Various physico-chemical properties such as solubility, viscosity, adhesive (shear) strength etc. of the in-house prepared adhesives are estimated. The cost-effective and eco-friendly composites are also prepared in the study using EPS and biomass (saw dust as a model compound), and the prepared composites are characterized as per the standard procedures. From the results, it is observed that MEK solvent offered maximum solubility of the EPS among the chosen solvents. The composite study revealed that post-consumer EPS and biomass materials have good candidate to the preparation of wood-plastic composites. More study is required to explore more sustainable applications from the considered recycled materials in the present study.

Acknowledgments

The authors express their gratitude to B V Raju Institute of Technology (BVRIT)-Narsapur, Medak District, Telangana, India for providing the necessary support for the present study.

References

Abdulkareem SA, Raji SA, Adeniyi AG, (2017). Development of particleboard from waste styrofoam and sawdust. Nigerian Journal of Technological Development, 14 (1): 18-22.

Appala Naidu U, Srikanta D, Bhanu Radhika G, (2020). Scientific and engineering aspects of potential applications of post-consumer (waste) expanded polystyrene: A review. Process Safety and Environmental Protection, 137:140-148.

Table 1. Solubility and viscosity values of the EPS derived adhesives.

Name of solvent	Adhesive code	Solubility at 30 ± 1 °C (wt.%)	Viscosity at 30 ± 1 °C (cP)
m-Xylene	A-X	40	4586
n-Butyl acetate	A-B	37	4206
Tetrahydrofuran	A-T	36	5934
Methyl ethyl ketone	A-M	42	6127
Gasoline	A-G	28	2296

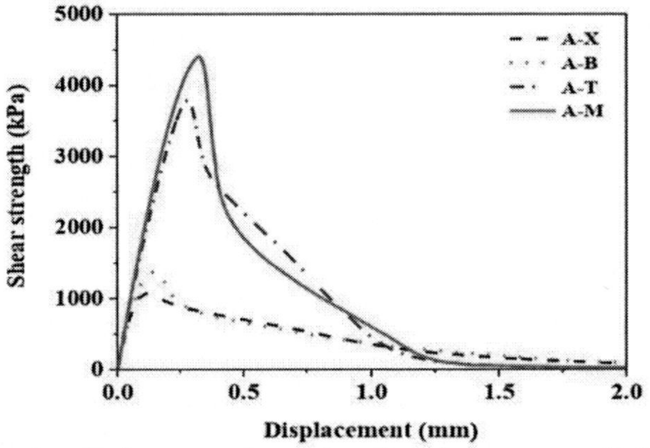

Figure 1. Shear strength analysis curves of wood-based specimens prepared with various adhesive.

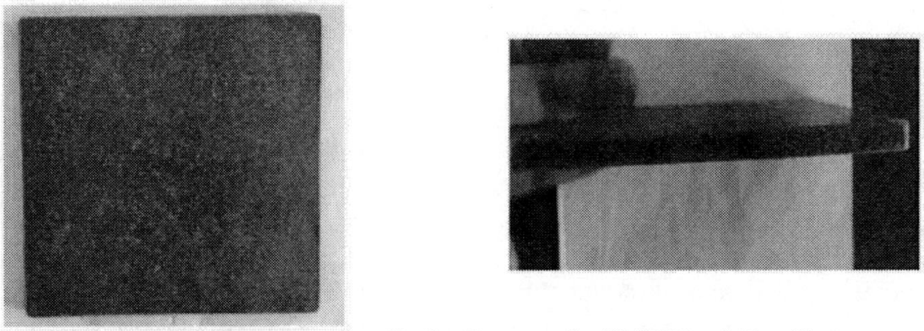

Figure 2. Wood-plastic composite sheet prepared with EPS and Saw dust.

Table 2. Effect of EPS quantity on the properties of the EPS-Biomass composite products.

EPS loading (wt.%)	Tensile strength (MPa)	Tensile modulus (MPa)	Elongation (%)	Flexural strength (MPa)	Water absorption (wt.%)
*20	-	-	-	-	-
*30	-	-	-	-	-
40	4.5	1450	0.72	16.4	7.9
50	19.3	3762	2.78	33.9	7.1

*Analysis is not done as the composite sheets are not formed properly.

248a Natural and Metal-Impregnated Clays Catalyze the Pyrolytic Treatment of Crude-Oil Contaminated Soils

Priscilla Dias da Silva, Sara Denison, Caroline Koester, Pedro J. J. Alvarez, Kyriacos Zygourakis

Environmental pollution by petroleum hydrocarbons poses serious threats to public and ecosystem health. About 98% of crude oil spills occur on land and exposure to terrestrial spills correlate to cancer, preterm births, neurocognitive deficits, and other public health concerns that can span across generations due to genetic or birth defects. Amongst the constituents of crude oil, polycyclic aromatic hydrocarbons (PAHs) are of particular concern since they are probable carcinogens and persist in the environment due to their recalcitrance to microbial degradation.

Soils impacted by relatively high concentrations of petroleum hydrocarbons are often remediated by thermal technologies that, while fast and effective, have high energy demands and destroy the fertility of treated soils [1]. Pyrolytic treatment of soils contaminated with crude oils is an alternative approach that offers significant advantages over other thermal methods. It can efficiently remove total petroleum hydrocarbons (TPH) to meet regulatory standards and detoxify the soil by removing toxic pollutants like PAHs with lower energy requirements than incineration [2,3]. Since it is carried out at lower temperatures than other methods, pyrolytic treatment can preserve soil components responsible for water and nutrient retention since restores (at least partially) the fertility of the soil, enhancing its value for ecosystem restoration and re-greening efforts [3,4].

In this study we show that clays impregnated with non-toxic transition metals (iron or copper) can be used as an amendment to decrease the required pyrolytic treatment temperature and time. Amending a weathered crude-oil contaminated soil with natural bentonite or bentonite modified with ion-exchanged Fe or Cu, achieved almost complete (95 to 99%) removal of residual total petroleum hydrocarbons (TPH) at a pyrolysis temperature of 370°C with 15-min treatment time. More importantly, pyrolytic treatment of amended soils at the unprecedentedly low pyrolysis temperature of 300°C achieved 87% TPH removal with Cu-bentonite and 93% with Fe-bentonite. Using TG-MS experimental data, we showed that the presence of transition metals significantly lowered the onset temperature of pyrolysis reactions. These results strongly support our hypothesis about the catalytic role of the transition metals and encourage further work to elucidate the reaction mechanisms and inform the design of optimized remediation processes. To determine the feasibility of this promising pyro-catalytic soil remediation approach, techno-economic analyses must be carried out to evaluate tradeoffs of increasing treated soil volume though modified clay amendments versus decreasing energy requirements and contact time.

References:

1. J. E. Vidonish, K. Zygourakis, et al, "Thermal Treatment of Hydrocarbon-Impacted Soils: A Review of Technology Innovation for Sustainable Remediation," *Engineering*, **2**, 426–437 (2016).

2. Y. Gao and K. Zygourakis, "Kinetic Study of the Pyrolytic Treatment of Petroleum Contaminated Soils," *Ind. Eng. Chem. Res.*, **58**, 10829–10843 (2019)

3. Y. Gao, P. Dias Da Silva, P. J. J. Alvarez, and K. Zygourakis, "Integrating Thermal Analysis and Reaction Modeling for Rational Design of Pyrolytic Processes to Remediate Soils Contaminated with Heavy Crude Oil," *Environ. Sci. Technol.*, **55**, 11987–11996 (2021).

4. W. Song, J. E. Vidonish, et al., "Pilot-Scale Pyrolytic Remediation of Crude-Oil-Contaminated Soil in a Continuously-Fed Reactor: Treatment Intensity Trade-Offs," *Environ. Sci. Technol.*, **53**, 2045–2053 (2019).

248b Abiotic Degradation of Chlorinated Solvents with Reactive Iron Minerals from Redox Transition Zones

Xin Yin, Han Hua, James Dyer, Richard Landis, Donna Fennell, Lisa Axe

Reactive iron (Fe) mineral coatings found in reduction-oxidation (redox) transition zones (RTZs) contribute to the attenuation of contaminants. This research focuses on the abiotic degradation processes of 1,4-dichlorobenzene (1,4-DCB), tetrachloroethylene (PCE), and trichloroethylene (TCE) with Fe mineral coatings from a contaminated site. An 18.3-m anoxic core was collected from the site, where constituents of concern (COCs) in groundwater included chlorinated solvents. In the earlier profile studies, RTZs were identified with a suite of analyses. Reactive Fe mineral coatings were abundant in RTZs, with distinct characteristics in surface composition, mineralogy, and morphology; sequential extraction was conducted for semi-quantifying speciation. In this research, batch studies were conducted to evaluate reaction kinetics with anoxic sediments bearing Fe(II) mineral nano-coatings spiked with either 1,4-DCB, PCE, or TCE. Control groups included pure pyrite and siderite. For 1,4-DCB treatment, although dechlorination was not observed over the time period of the study in the control groups, reaction kinetics with RTZ sediments followed second order rate expressions. Chlorobenzene and benzene were detected as byproducts, suggesting hydrogenolysis reduction. The second-order rate constants for the Fe(II) mineral nano-coatings in 1,4-DCB degradation resulted in $1.24'10^{-3}$ L g^{-1} $h^{-1}$for mackinawite (FeS), $1.73'10^{-3}$ L g^{-1} h^{-1} for pyrite (FeS_2), $1.89'10^{-4}$ L g^{-1} h^{-1} for siderite ($FeCO_3$), and $1.79'10^{-4}$ L g^{-1} h^{-1} for magnetite (Fe_3O_4). The high reactivity of Fe mineral coatings expects to be due to large surface areas. PCE and TCE reduction were observed in the control and sediment groups, which also followed second-order rate expressions. For these three chlorinated solvents, the trend for dechlorination activity of Fe(II) mineral nano-coatings followed: Fe(II) sulfide minerals > magnetite > siderite. As a result, reactive Fe mineral nano-coatings are expected to play a significant role in the attenuation of chlorinated solvents in contaminated subsurface environments.

248c Chlorine Dioxide Decay in the Presence of Lead Minerals Found in Drinking Water Distribution Systems

Christian Lastoskie

Chlorine dioxide (ClO_2), an alternative disinfectant to free chlorine, has been used to reduce the formation of organic disinfection byproducts during water treatment. ClO_2 is highly reactive, and its decay is enhanced in the presence of lead oxide (PbO_2), lead carbonate ($PbCO_3$), and cupric oxide (CuO), minerals commonly found in the corrosion scale on distribution system pipes. Reaction of these corrosion minerals with chlorine dioxide and the byproducts created during said reactions were investigated. In addition to corrosion mineral, initial concentration of ClO_2 and pH were both varied. Reaction took place in a batch reactor, ClO_2 was measured via ultraviolet-visible spectrophotometry, and byproducts were quantified using ion chromatography. Only chlorite is produced from ClO_2 decay when lead minerals are present, whereas both chlorite and chlorate are produced by disproportionation when CuO is present. ClO_2 decay occurs at a significantly higher rate on PbO_2 compared to CuO. The reaction rate for ClO_2 decay, in the presence of PbO_2, peaks at pH 8.3, which corresponds to the point of zero charge of PbO_2. This indicates that surface charge may play a significant role in lead mineral-assisted decay reactions with disinfectants, including ClO_2. The reaction rate dependence for ClO_2 decay on PbO_2 is well described by a pseudo-second-order adsorption kinetic model. These findings highlight the need for chlorite, but not chlorate, byproduct management when ClO_2 is used as a disinfectant in water distribution systems with lead-based pipes. Robust strategies for management of a chlorine dioxide residual must be implemented in both copper and lead-based service lines to ensure a disinfectant persists and protects against microbial contamination in premise plumbing.

248e Design of Porous Hybrid Composites with Dual-Adsorption Capacity for Carbon Capture

Vijay T. John, Oluwole Ajumobi, Azeem Farinmade, Julia A. Valla

Metal organic frameworks (MOFs) have gained significant recognition in their applications for capture due to their capability to physically capture CO_2 molecules within their pore structure. They possess high surface area, and their surface can be functionalized with amines to increase their efficiency. However, amine functionalization leads to the binding of the amine group to the open metal sites which is responsible for carbon capture via physisorption and thus, adsorption of CO_2 occurs through chemisorption over the bonded amine groups. In this study, we encapsulated HKUST-1 crystals in porous silica spheres using a one-step facile aerosol-assisted synthesis technique. This ship-in-a-bottle approach is such that, as the silica growth occurs in the heating zone due to evaporation induced hydrolysis and condensation, the HKUST-1 crystals are entrained within the porous spherical matrix. Characterization results shows that the structural properties and crystallinity of the HKUST-1 particles remained unaltered within the porous shell, after encapsulation. We showed the accessibility of the entrained HKUST-1 to CO_2 capture at atmospheric pressure and 35 °C at dry conditions, using a thermogravimetric analysis setup. The results showed that the HKUST-1 crystals are fully accessible to CO_2 molecules with an adsorption capacity of 1.78 mmol/g at 30 wt% of HKUST-1 (PS30HKT). When normalized, the PS30HKT has an adsorption capacity of 5.93 mmol/g which is comparable to that of bulk HKUST-1 (reference sample) with CO_2 adsorption capacity of 5.62 mmol/g (wt%). This shows that the encapsulated HKUST microcrystals are fully accessible to CO_2 through the narrow pores of the PS30HKT without loss in adsorption performance. This unique composite morphology has potential for further functionalization to improve the CO_2 capture of the composite material such that, the porous silica matrix can be functionalized with amine groups while the surface and pore structure of the encapsulated HKUST-1 crystals remained pristine. In essence, we will have a hybrid adsorbent that can capture CO_2 via physisorption over the entrained HKUST-1 crystals while the amine-functionalized silica pores can simultaneously adsorb CO_2 via chemisorption.

248f On-Demand Photocatalyzed Flow Production of Hydrogen Peroxide from Immobilized Anthraquinones

Hunter Vibbert, Christoph Bendel, Mengqi Shen, Ah-Hyung Alissa Park, Aaron Moment

Aqueous hydrogen peroxide is an important, versatile commodity chemical across many concentrations ranges that is used widely in the treatment of wastewater, in bleaching applications, as an environmentally friendly disinfectant, and a reagent in chemical technologies. Conventionally, H_2O_2 is produced at scale thermally via the anthraquinone oxidation process (AO) using a combination of molecular anthraquinones, heterogeneous Pd catalysts, and H_2. A process that could eliminate the need for Pd would have environmental, separation, and process intensification advantages. We have previously shown that a biphasic system containing anthraquinone dissolved in organic solvent and water could produce commercially relevant concentrations of H_2O_2 (e.g., 2% w/w) in under 3h. Here, we discuss the development of a technology based on immobilized anthraquinones that can be directly excited with light to produce H_2O_2 in the presence of a hydrogen carrier (e.g., organic solvent or water) batchwise or in a flow configuration. The design of the flow reactors for H_2O_2 production, characterization of the immobilized anthraquinone catalysts, and their effect on H_2O_2 production will be discussed.

18a Microbial Soil Amendments for Modulating Soil Moisture Dynamics for Improved Water Availability in Agriculturally Relevant Soils

Moises M. Gutierrez, Micah V. Cameron-Harp, Partha P. Chakraborty, Emily M. Stallbaumer-Cyr, Jordan A. Morrow, Ryan Hansen, Melanie M. Derby

The world is changing; the global population is increasing rapidly, and climate change has caused the food, energy, and water nexus to become increasingly complex and interdependent. In particular, the growing global demand for food combined with frequent periods of drought have stressed agriculture systems in semi-arid regions, such as western Kansas. A potential solution to water scarcity focuses on improving soil water retention during a precipitation or irrigation event by increasing water holding capacities using soil inoculants. This work investigates the application of *Bacillus subtilis* and its bioproduct, Surfactin, to break down soil water repellency and enhance water holding capacity, ultimately reducing irrigation water demands in loam and sandy loam soils. Laboratory-scale experimental results show that both *B. subtilis* and Surfactin increase the evaporation rates when soil moisture is abundant up to 43% relative to the controls and decrease the evaporation rates when moisture is scarce up to 49% relative to the control samples. A System Dynamics Model (SDM) is used to evaluate the potential significance of these altered soil moisture dynamics in agriculturally-relevant growing conditions found in western Kansas. The combined laboratory-scale experiments and SDM results project a significant decrease in the soil moisture threshold for treated soils under typical precipitation patterns. As a result, the Kernel density approximation for each treatment suggests an increase in the number of days with evaporation rates at higher soil moisture on a 100-day growing season simulation. Overall, these soil amendments could deliver a benefit to stressed agriculture systems by increasing soil moisture availability due to the prolonging of the evaporation rates.

18b An Energy Efficient Nitrogen Removal Process Via Simultaneous Nitrification and Denitrification in Wastewater

Qingkun Wang, Jianzhong He

Bacteria capable of simultaneous nitrification and denitrification (SND) and phosphate removal could eliminate the need for separate reactors to remove nutrients from wastewater and alleviate competition for carbon sources between different heterotrophs in wastewater treatment plants (WWTPs). Here we report a newly isolated *Thauera* sp. strain SND5, that removes nitrogen and phosphorus from wastewater via SND and denitrifying-phosphate accumulation, respectively, without accumulation of metabolic intermediates. Strain SND5 simultaneously removes ammonium, nitrite, and nitrate at an average rate of 2.85, 1.98, and 2.42 mg-N/L/h, respectively. Batch testing, detection of functional genes, nitrogenous gas detection and thermodynamic analysis suggested that nitrogen gas, with hydroxylamine produced as an intermediate, was the most likely end products of heterotrophic ammonium oxidation by strain SND5. The generated end products and intermediates suggest a novel nitrogen removal mechanism for heterotrophic ammonium oxidation in strain SND5 (). The heterotrophic ammonium oxidation to nitrogen gas without emission of N_2O by strain SND5 is a particular advantage from the perspective of greenhouse gas emissions in WWTPs. As a greenhouse gas, N_2O shows 298 times the atmospheric heat-trapping capacity of CO_2, and its emission from WWTPs is associated with denitrification of nitrate and nitrite during biological nutrients removal. Since the majority of NO_2^- and NO_3^- in WWTPs is derived from ammonium oxidation, bypassing this pathway could have a major impact on overall N_2O production. Since strain SND5 generates neither NO_2^- or N_2O during heterotrophic ammonium oxidation, its metabolic activity could help to limit greenhouse emissions from WWTPs while at the same time leads to energy saving.

18c Nutrient Recovery from Hydrothermally-Treated Dairy Manure Using Algae Cultivation

Callan Glover, Nicholas Silva, Gabriela Quiles, Pablo Cornejo, Sage Hiibel

Increased demand for dairy products has led to intensification of farms, resulting in significant environmental impacts. Manure management is one of the most significant contributors to environmental impacts through greenhouse gas emissions (GHGs) and contamination of water resources by nitrogen (N) and phosphorus (P). As such, innovative treatment methods are needed to handle large quantities of manure and mitigate the environmental impacts within this food-energy-water system. Recent work has focused on energy and nutrient recovery of the organic matter and nutrients found in manure. Hydrothermal carbonization (HTC) is an emerging thermochemical treatment that converts organic waste into a carbon-rich solid fuel and a nutrient-rich aqueous product (HAP). The solid fuel can be used for energy production as a substitute for fossil coal. However, the HAP requires further treatment to capture nutrients and mitigate environmental impacts. In this research, algae species commonly used for nutrient removal in wastewater are evaluated for cultivation in HAP: *Spirulina maxima, Chlorella vulgaris, Scendesmus obliquus,* and *Chlamydomonas reinhardtii*. Species performance is quantified using growth rate, total biomass production, and removal of nitrate, ammonia, total nitrogen, and total phosphorus from various concentrations of HAP. Preliminary results show all species are capable of growth in HAP at low (<10%) concentrations and *S. maxima* shows high nutrient removal (>94% total nitrogen and total phosphorus removed) in low concentrations of HAP, improving water quality of the effluent. In addition, microalgae biomass is rich in protein and can be utilized as a food source for dairy cattle, directly recycling nutrients on-farm. Integration of algae production with HTC can help recover valuable nutrients from manure and improve water quality, alongside renewable energy production.

18e Phytoremediation of Deicing Salt from Roadside Soils

Leif van Lierop, Bo Hu

Rock salt, consisting of primarily sodium chloride (NaCl), is a cheap and effective deicing agent for winter weather events in northern climates. However, most of the salt dissolves in melting snow and ice and eventually finds its way into soil, streams, lakes, wetlands and groundwater. The Minnesota Pollution Control Agency (MPCA) found that 39 of 340 lakes tested for chloride were above the chronic exposure limit that has been shown to be lethal to aquatic life. Sodium accumulates in soil where it can negatively impact soil structure and plant growth. Contaminated soil can be up to 10 meters from the side of the road. Sodium concentrations of roadside soils measured in the Twin Cities Metro Area ranged from around 150 ppm to exceptionally high levels of roughly 8200 ppm. Many forms of soil remediation are costly and disruptive to the soil. Phytoremediation, on the other hand, uses salt tolerant plants to uptake salt into their biomass, which is then harvested. Although phytoremediation can require multiple seasons or harvests to remediate a site, it is an inexpensive and environmentally non-disruptive method. Also, the harvested plant biomass may provide further uses in industrial applications like serving as animal feed or sources of energy.

A species inventory of potential plant species was created to determine the most promising species for salt phytoremediation of roadside soil. Greenhouse experiments were then performed with the highest rated species, including common sunflower (*Helianthus annuus*), pitseed goosefoot (*Chenopodium berlandieri*), and big bluestem (*Andropogon gerardii*). Plants were grown for roughly 7 weeks in 10 cm diameter pots with 75 g of fresh Sungrow potting soil and watered weekly with 100 mL of 0, 50, or 100 mM NaCl saltwater solution for a total of 4 weeks during the experiment. Common sunflower showed the highest total salt uptake into aboveground biomass with 51.4 and 58.5 mg Na^+ and 162.4 and 185.0 mg Cl^- for the 50 and 100 mM NaCl treatments respectively. Pitseed goosefoot accumulated 19.6 and 29.4 mg Na^+ and 89.9 and 101.3 mg Cl^- while big bluestem showed the lowest salt uptake with only 2.92 and 1.65 mg Na^+ and 23.6 and 9.36 mg Cl^- for the 50 and 100 mM treatments. According to the amount of salt added in the 50 mM treatment, common sunflower removed 17.8% and pitseed goosefoot removed 9.15% of the total salt from the soil. Therefore, common sunflower could theoretically remove most of the salt within 5 to 6 harvests and it would take around 10 to 11 harvests for pitseed goosefoot. These results show that common sunflower and pitseed goosefoot may be great candidates for salt remediation on salt contaminated roadsides and should be further explored with field studies on roadsides contaminated with deicing salt.

Keywords: phytoremediation, road salt, deicing salt

Caffeine Removal Using "The Miracle Tree" Wastes For Sustainable Water Purification Processes As Alternative To Classic AOP And Commonly Used Filtration Media

Andrea C. Landázuri[1,2,3*], Alina L. Trávez[1], Israel Sornoza[1], Nicole González[1], David A. Egas[1], Lourdes Orejuela-Escobar[1,2,3], Rodny Peñafiel[4]

[1]Universidad San Francisco de Quito USFQ, Colegio de Ciencias e Ingenierías; Applied Circular Engineering & Simulation Group (GICAS), Departamento de Ingeniería Química, *Diego de Robles y Vía Interoceánica, P.O. Box 17-0901, Quito, Ecuador.*
[2]Universidad San Francisco de Quito USFQ, Instituto de Investigaciones Biológicas y Ambientales BIÓSFERA, *Diego de Robles y Vía Interoceánica, P.O. Box 17-0901, Quito, Ecuador.*
[3]Universidad San Francisco de Quito USFQ, Instituto iBioMed, Diego de Robles y Vía Interoceánica, P.O. Box 17-0901, Quito, Ecuador.
[4]Universidad Técnica de Ambato UTA, Facultad de Ciencia e Ingeniería en Alimentos y Biotecnología, Campus Académico Huachi – Río Payamino y Av. Los Chasquis, Ambato, *Ecuador*

*Corresponding author: alandazuri@usfq.edu.ec

Abstract

This work showcases the use of *Moringa oleifera* Lam. (MO) residues for the removal of a selected polar emerging organic compound (EC), caffeine, from water. The mechanisms of coagulation-flocculation, degradation, and adsorption are discussed by comparing green alternatives that uses MO cotyledon (MOC) extracts against Advanced Oxidation Processes (AOPs), as well as filtration experiments with moringa husk and commonly used materials (zeolite, activated carbon, gravel). Relevant coagulation-flocculation results show that caffeine removal using MOC extracts provide statistically similar results compared to a Fenton process. In addition, dynamic results found that the medium that removed most caffeine per unit mass was that of *Moringa oleifera* Lam. husk which is described by a physisorption process that follows the behavior of a deduced model (Bulk Balance Filtration Model-BBFM) that depends mainly of operation time and filter depth. The use of MO lignocellulosic wastes from agroindustry reduces the negative impact of these wastes in the environment and allows and efficient and sustainable use of a natural renewable resource, which otherwise does not enter in the value chain of the agroindustry production; moreover, carbon foot print may be reduced and communities may live in healthier living conditions.

Key words: Caffeine, water purification, AOP, filtration, *Moringa oleifera* Lam.

1. Introduction

Advanced oxidation processes (AOPs) and slow sand filtration are traditional methods to remove emerging organic contaminants (ECs) from water. AOPs involve reactions in which hydroxyl radicals (•OH) are produced in water [1]. The well-known Fenton reaction that combines H_2O_2 and Fe^{2+}, to oxidize Fe^{2+} to Fe^{3+} and generate a hydroxyl radical [2]–[4]. In the presence of an organic substrate (RH), the hydroxyl radical abstracts a hydrogen atom, generating an organic radical (R•), which undergoes a transformation resulting in oxidation products. An excessive concentration of Fe^{2+} and H_2O_2 can convert organic compounds to CO_2 and water. However, the unspecific reactivity of •OH towards organic substrates may limit AOP applications to remove particular pollutants in complex wastewater matrices [5]. Furthermore, Fenton and Photo-Fenton reactions are largely restricted to acidic pH ($pH \leq 3$) since Fe^{3+} ions precipitate as hydroxide at higher pH [6]. Ferro-peroxy complexes and iron ions generate the radical species that initiate the chain reactions leading to the degradation of caffeine.

Conversely, alternative methods to AOPs are proposed considering coagulant/flocculant and adsorbent materials. There is evidence suggesting that the Moringa's seed extract has an affinity to organic compounds reflected in the adsorption of Carmine-Indigo Dye [7] and removal of tetracycline antibiotic [8]. The seeds have served as a coagulant in the clarification of drinking water and for treatment in wastewater due to the presence of a cationic coagulation protein soluble in water, which has the ability to reduce turbidity [9]–[13]. Due to its water solubility, caffeine is considered a model molecule for all kinds of polar organic compounds to be registered as secondary contaminants [14]. Caffeine may also serve as a model for other types of ECs, which can help understand and investigate their behavior against different treatments.

Along with the *Moringa oleifera* plant, also known as "the miracle tree", there are other candidates that may be used for emerging organic pollutants, including activated carbon [15], zeolite [16] and pulverized construction rocks which is most often referred to as gravel. Most of these materials appear to remove organic compounds by adsorption phenomena thanks to Van der Waals forces (either induced, permanent or transient electric dipoles) between contaminants and adsorbents and materials that experience physisorption may be used in depth filtration, as they can be cleaned and reused several times [17]. Hence, these can be incorporated in slow sand filters as an efficient way to remove the ECs of interest.

Unlike other treatments for ECs such as Advanced Oxidation Processes (AOPs) like Fenton, photo-Fenton and photocatalysis, slow sand filters require very little mechanical power, chemical or replaceable parts to work [18]. Examples of filtration models are the: Bohart-Adams [19], Thomas [20], [21], Yoon-Nelson [22], and Clark [23]; while a most recent model is the Bulk Balance Filtration Model (BBFM) [24] obtained by combining Iwasaki and Ives equations, along with a mass balance around the filtration system. It is important to investigate the contaminant removal mechanism that takes place inside the filter. Different filter beds may have different models that adapt to specific cases, which is why every medium needs to be studied independently before a combination is done. The first variable of interest that needs to be examined is caffeine concentration (dependent variable) as a function of operation time and filter bed depth (independent variables). There are numerous other control variables that should be taken into consideration such as temperature, packaging effectiveness, and water flow; however, most of them are implicit in the breakthrough curve model's parameters that are utilized [23].

Efforts to reduce pollution in the rivers surrounding Quito, the capital of Ecuador, have been done by the Public Metropolitan Drinking Water and Sanitation Enterprise (EPMAPS) through a program that aims to manage wastewater treatment and discharge, in compliance with current environmental regulations in the city of Quito [25]. Nevertheless, the project does not consider ECs treatment. Caffeine is one of the most abundant EC found in a main wastewater discharge of Quito with a concentration around 6 [mg/L] [26]. In addition, MO is being cultivated and harvested in the Ecuadorian coast, and MO residues such as husk and pressed cotyledons are currently being discarded as waste and even though it is not used in any food industry, it has been proved that it can be applied for adsorption removal of ECs as already mentioned.

In this work batch experiments with several initial concentrations of caffeine (5, 25, 50 and 100 ppm) in water were submitted to the following removal processes: Fenton, photo-Fenton, modified Fenton, coagulation/flocculation using moringa oleifera cotyledon (MOC) extracts; furthermore, filtration experiments involving a dynamic analysis of depth filters using caffeine as the model contaminant and *Moringa oleifera* Lam. husk, zeolite, activated carbon, and gravel as filtering beds. Filtration media were used without any further treatment and the selected filtration column is a cylinder-shaped borosilicate filter with 11 mm diameter and 60 cm height [24], [27]. Ultimately, the objective of this investigation is to provide insightful information regarding different, simple, practical technologies for caffeine removal with the perspective of supporting future scaling-up purposes while promoting the use of residual lignocellulosic materials for green water treatment in Ecuador and overseas.

2. Materials and methods

2.1 Batch Experiments

Two sets of batch experiments were performed. Set one involved the treatment of synthetic wastewater samples at a starting initial concentration of 100 ppm subjected to the following processes: Fenton and photo-Fenton, photocatalysis using TiO_2, and coagulation/flocculation using MOC extracts along with NaCl (MOC1-NaCl). The second set of experiments explored the purification of caffeine solutions at lower initial concentrations (5, 25 and 50 ppm), which were treated through a modified Fenton process (an optimized version) and MOC with NaCl (MOC2-NaCl) with slight variations from the 100 ppm experiments. MOC2-NaCl plus caffeine experiments were designed in order to reflect a more realistic situation in terms of the abundance of caffeine in one of the most important discharges of Quito as well as to propose dosages (in mL) of the MO extracts into the caffeine samples which provides a more simple and practical way compared to MOC1-NaCl and caffeine experiments.

2.1.1 Materials and reagents

Moringa oleifera Lam. seeds (Ecuamoringa, Guayaquil-Ecuador) and NaCl 99.5% (Loba Chemie) were used for MOC-NaCl extraction experiments. Artificial water was prepared with caffeine standard (Merck). Assays for the 100-ppm caffeine initial concentration experiments involved caffeine 99 wt% (Aldrich), TiO_2 P25 from Degussa AG (Germany), H_2O_2 33 wt% (Merck), ferrous sulfate heptahydrate 99 wt% (Aldrich), sodium thiosulfate 99.5 wt% (Aldrich). Lower concentrations assays (5-50 ppm) used 50 wt% H_2O_2 (Merck), 99 wt% ferrous sulfate heptahydrate (Sigma-Aldrich), and 98% H_2SO_4 (Merck). HPLC grade methanol (Fisher Scientific), and glacial acetic acid (Fisher Scientific) were used as received without any further purification. E-Pure-grade water was produced in the laboratory and used to prepare the solutions in all experiments.

2.1.2 Assays for caffeine initial concentration of 100 ppm

Dilutions were made to give rise to a 100 μg/mL caffeine stock solution. Serial dilutions were performed in 100 mL flasks to prepare 5, 10, 50, 100 μg/mL solutions used later to obtain the calibration curve. Standards were degassed for 30 minutes with ultrasonic cleaner (model 5510R-DTH) prior injection. Approximately 3 mL of each solution were filtered with nylon syringe filters (Millipore, Millex-HV) and placed in eppendorf tubes. Five-level caffeine Calibration curves were used prior samples injection.

For Fenton and Photo-Fenton experiments, a 100 mL caffeine aqueous solution (0.52 mmol/L) was used. The pH of the solution was adjusted to 2.9±0.2 with sulfuric acid (99%, 0.1 M). Ferrous sulfate (1-10 mg, 0.036-0.36 mmol/L final concentration) was used, and 0.5 mL 33% m/m hydrogen peroxide solution (5.4mmol/L final concentration) were added under constant stirring Fenton process was carried out under these acidic conditions at room temperature in a 250 mL beaker.

For the photo-Fenton process, a home-made wood reactor equipped with a UV-lamp (352nm, 20W) and three ventilators (12V each) was used. The caffeine solution was allowed to sit in the center of the reactor under stirring conditions [3], and was irradiated with light. 1.5 mL aliquots were withdrawn from the aqueous solution at different times over a one-hour period. Both Fenton and Photo-Fenton reactions were stopped with an excess of sodium thiosulfate. Aliquots were filtered using nylon filters and protected from light for approximately 12 hours in a 4 °C refrigerator prior to HPLC analysis.

For the photocatalytic experiments with TiO_2, the pH of a 100 mL caffeine aqueous solution (0.53mmol/L) was adjusted to 7.0. and 0.1 g of TiO_2 (12.5mmol/L final concentration) was added to the aqueous solution. The solution was stirred and allowed to sit without light for approximately 30 minutes. The solution was then irradiated by a UV-lamp (352 nm) using the same reactor used for Photo-Fenton samples. Then, 1.5-mL aliquots were withdrawn from the aqueous solution at different times for a one-hour period and then centrifuged at 3000 rpm for 10 minutes. Samples were filtered with nylon filters and protected from light for approximately 12 hours in a 4 °C refrigerator prior to HPLC analysis.

Caffeine samples provided for HPLC analysis, were treated with NaCl 1 M. For this, MO seeds were thoroughly washed and dried at 60°C overnight. The seeds were then crushed and weighted. 1-5 grams of the treated seed were added to100mL of an artificial caffeine solution (100ppm caffeine final concentration) containing NaCl 1 M. Solution was vigorously stirred at room temperature for 45 minutes. Then, 1-1.5mL aliquots were withdrawn from the aqueous caffeine-cotyledon extract solution at different times, filtered, and protected from light in a 4 °C refrigerator prior to HPLC analysis. The resulting extract is referred to as the MOC1-NaCl extract.

2.1.3 Assays for caffeine initial concentrations between 5-25 ppm

A modified Fenton process was followed. For this, artificial caffeine 100-mL stock solutions were prepared at 5, 25 and 50 µg/mL. Next, the pH of each one was adjusted with 0.1 M sulfuric acid until achieving 2.6 ± 0.2 pH values at 25 °C [28]. To start Fenton reactions, 0.28 mL of hydrogen peroxide and 5 mg of heptahydrate iron sulphate were added. Experiments were performed in triplicate.

Aliquots were taken at 0, 5, 10, 15, 30, 45 and 60 min with a syringe and passed through a 0.45 µm filter, placing them in 1.5-mL Eppendorf tubes. Additionally, 0.5 mL of sodium thiosulfate (2.64×10^{-2} M) were added in order to stop the reaction. All samples were kept at a temperature of 4 °C until further HPLC analysis.

To obtain the so-called MOC2-NaCl extract, 50 g of complete MO seeds were weighed and washed six consecutive times with Type 2 water; the clean seeds were then placed on an aluminum tray, in a single layer, and dried in an oven at a constant temperature of 60 °C, for 24 hours. The seeds were removed from the oven and placed in a desiccator for 30 min, or until reaching room temperature. The MO seeds dry-mass was then weighed. MO husks were removed leaving the cotyledons, only. The cotyledons were crushed with a mortar and pistil. Crushed MO cotyledons (MOC) were passed through a sieve, selecting the most abundant fraction (500-710 µm), only 5 g of MOC were placed in a volumetric flask until filled to a 100-mL volume with a 1M NaCl solution. This mixture was submitted to a batch process with magnetic stirring for one hour at 300 rpm and a temperature of 20 °C [7]. This extract was filtered in three different steps: using a coffee filter, then a H&R brand filter paper and a 0.45-µm filter. The resulting extract is referred to as the MOC2-NaCl extract. Several dosages of MOC2-NaCl between 1-10 mL were added to the caffeine samples and stirred at 400 rpm and room temperature while monitoring pH.

High-performance liquid chromatography HPLC analyses were conducted using a Buck Scientific BLC-10/11 system equipped with a manual injection valve with a 20 µL (for caffeine initial concentration of 100 ppm) and 50 µL (for initial concentrations of 5-50 ppm) sample loops and a fixed-wavelength UV detector set at 254 nm. An isocratic pump was used to deliver the eluent at a constant rate of 1 mL/min. The detector was set for 10-mV output, and resolution was set to 0.5 and 0.1, for experiments involving initial concentrations of caffeine of 100 ppm and 5-50 ppm, respectively.

A Pinnacle II C18 (5-µm particle size, 150 mm x 4.6 mm i.d.) reversed-phase (RP) analytical column with no guard column was used at room temperature. The mobile phase used was a solution of CH_3OH:CH_3COOH:H_2O (15:1:34) v/v/v. The mobile phase was mixed and filtered with PVDF 0.45 µm (Millipore, Millex-HV) nylon filter membrane, and degassed for 30 minutes with ultrasonic cleaner (model 5510R-DTH) before use. A PeakSimple (4.26 version, 6 channels) integrator was operated to record chromatograms and peak areas. The computer software used was PeakSimple (4.17 version).

Ultraviolet (UV) absorbance measurements were performed using a UV-VIS Cecil (model CE2041) equipped with quartz cell (1 cm path length). Absorbance was measured with water as the blank. The scan rate used was 300 nm/min with a data-point interval of 0.5 nm, in a 200-400 nm wavelength range.

Results were verified with a one-way analysis of variance (ANOVA) to determine if the treatments were statistically different or not. Also, the Tukey multiple comparison procedure was applied. This tool demonstrates if the variances are similar, furthermore, it considers all the simultaneous comparisons of the factors under analysis [29].

2.2 Dynamic Filtration experiments

The performance of four filtration media: granular activated carbon, zeolite, gravel and moringa husk, was analyzed. With regards to moringa, the larger husks (with approximate dimensions of 0.57 cm x 0.45 cm, and 0.10 cm were preferred for filling up the filtration column, rather than the small ones (i.e. broken husks). The filtration column consisted of a borosilicate column with 11 mm inner diameter and 60 cm height. The complete flowrate calibration and experimental procedures (caffeine concentration as a function of depth and time) are explained in previous works [24], [27].The complete system operated at a flowrate of 13.84 mL/min, with an initial caffeine concentration of 50 ppm, and the selected filtration bed heights (depths) varied from 13 cm to 54 cm, depending on the filtration media. Experiments as a function of time had a duration of 6 hours.

For each filtration media caffeine concentration was quantified using HPLC or UV-VIS spectroscopy (CECIL 2021 along with a 10 mm quartz cuvette) [27]. A stock solution of 1000 ppm caffeine (Sigma-Aldrich Caffeine Powder) was prepared, and subsequently caffeine solutions with specific concentrations (0, 5, 15, 30, 50 and 60 ppm). In the case of UV-VIS spectroscopy, quantification was done measuring the absorbance for each sample at a wavelength of 273.5 nm. When using HPLC, both, the height and the area below the characteristic peak of caffeine were measured for each concentration.

3. Results and Discussion

3.1 MOC1-NaCl and 100-ppm caffeine concentration

In our study we have analyzed different methods for caffeine removal. These processes vary in their nature of contaminants removal. On one hand, Fenton and Photo-Fenton processes involve caffeine degradation, whereas the use of Moringa seed extracts involves an adsorption process. These different procedures were all analyzed by HPLC-UV to provide insights on their efficiency. The caffeine concentration was chosen to be 100 ppm, as a starting point to simulate pharmaceuticals concentrations present in wastewaters.

For Fenton-like processes, pH has been found to have a crucial role in the oxidation efficiency, and maximum degradation has been observed at pH 2-4 [3]. Therefore, we chose pH 3 for our experiments. Caffeine degradation is known to be carried out by H_2O_2/Fe, which molar ratio has been vastly studied [30], and was not the focus of our study. However, we did vary the amount of Fe(II), as it was deemed to be an important factor in previous studies [31]. It was found that the larger the amount of Fe(II), the higher the degradation of caffeine. However, when 0.18 mmol/L and 0.36mmol/L of ferrous is used similar caffeine degradation is observed (see Fig. 1(a)). Moreover, in Fig. 1(b), we observe that Photo-Fenton produces a higher caffeine degradation (99%) when compared to Fenton (66%) at the same Fe(II) concentration (0.36mmol/L). This leads to the conclusion that the use of a light source increased the contaminant degradation significantly [32], and is therefore a more efficient method for caffeine degradation than the Fenton-only process. In the Photo-Catalysis family, we also find another commonly used method. This involves the use of TiO_2 /UV, which has been used to explore caffeine degradation [33]. For our study, we used TiO_2 (0.1g, 12.5mmol/L final concentration), and observed a lower caffeine degradation (53%) compared to the other Photo-Catalysis methods employed (see Fig. 1(b)). However, TiO_2 process conditions could be optimized in the future to yield to an even higher caffeine degradation, comparable to the one observed for Photo-Fenton.

On the other hand, we find that a practical way to remove organic contaminants, although through a very different process than the one observed in Fenton and Photo-Fenton methods, is the use of *Moringa Oleifera* seeds extracts. These seeds are known to have flocculant functionalities [34]. Therefore, caffeine is not degraded but rather removed through adsorption and coagulation-flocculation mechanisms by the seeds extract. This process does not involve the use of any chemicals nor light chambers, and constitutes a simple procedure that can be easily applied and extended to other compounds.

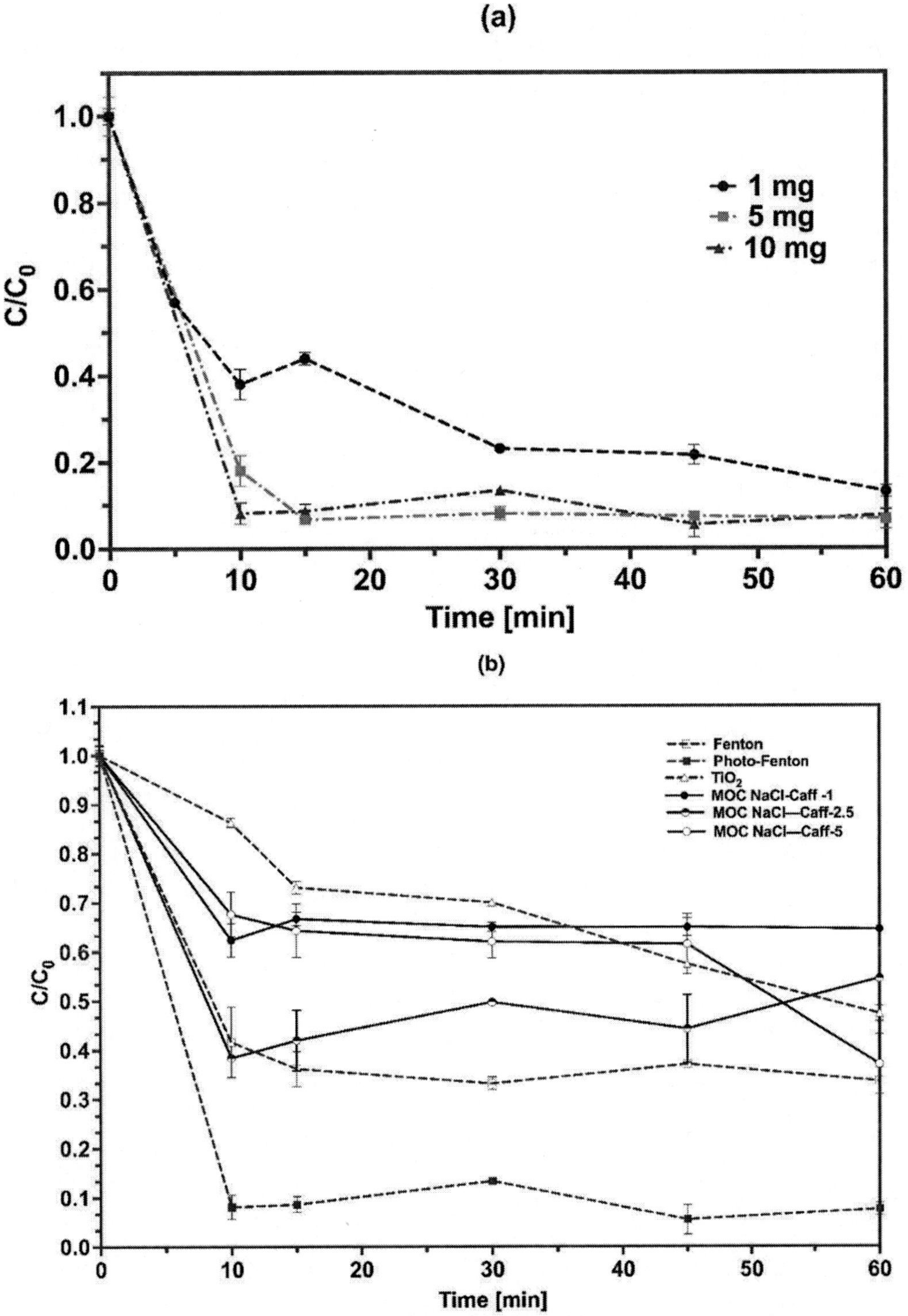

Fig. 1. AOPs results for the 100-ppm experiments set. (a) Photo-Fenton Process. Higher caffeine removal (up to 99%) was achieved with higher concentrations of Fe (10 mg, 0.36mmol/L); (b) AOPs comparison: TiO2 (0.1g, 12.5mmol/L) yielded a 53% caffeine removal; Fenton with Fe (10mg, 0.36mmol/L), 66%; and, photo-Fenton (with 10 mg of Fe, 0.36 mmol/L), 99%. Removal results using 1, 2.5, and 5 g of MOC-NaCl is included to illustrate the similar trend when compared to Fenton processes.

On Fig. 1(b), we further compare the different caffeine removal methods used. It was found that moringa oleifera yielded a higher caffeine removal when compared to TiO_2, but a lower removal when compared to the other AOPs. However, due to the simplicity of the MO procedure, it can be easily used as a method for wastewater treatment. Among the different amounts of MOC-NaCl used, higher caffeine adsorption is observed with a bigger amount of Moringa seeds (5g), but at lower amounts (2.5g) the adsorption is observed to happen faster than what is observed with 5g. It is also observed, that with the lowest amount of Moringa seeds (1g) the adsorption process reaches a plateau after 15 minutes, whereas for 5g, further adsorption takes place even after 45 minutes (Fig. 1(b)). Therefore, 2.5 g of seeds extract produce a faster caffeine adsorption than 1g, but 5 g, even though slower, yield a higher final caffeine adsorption.,

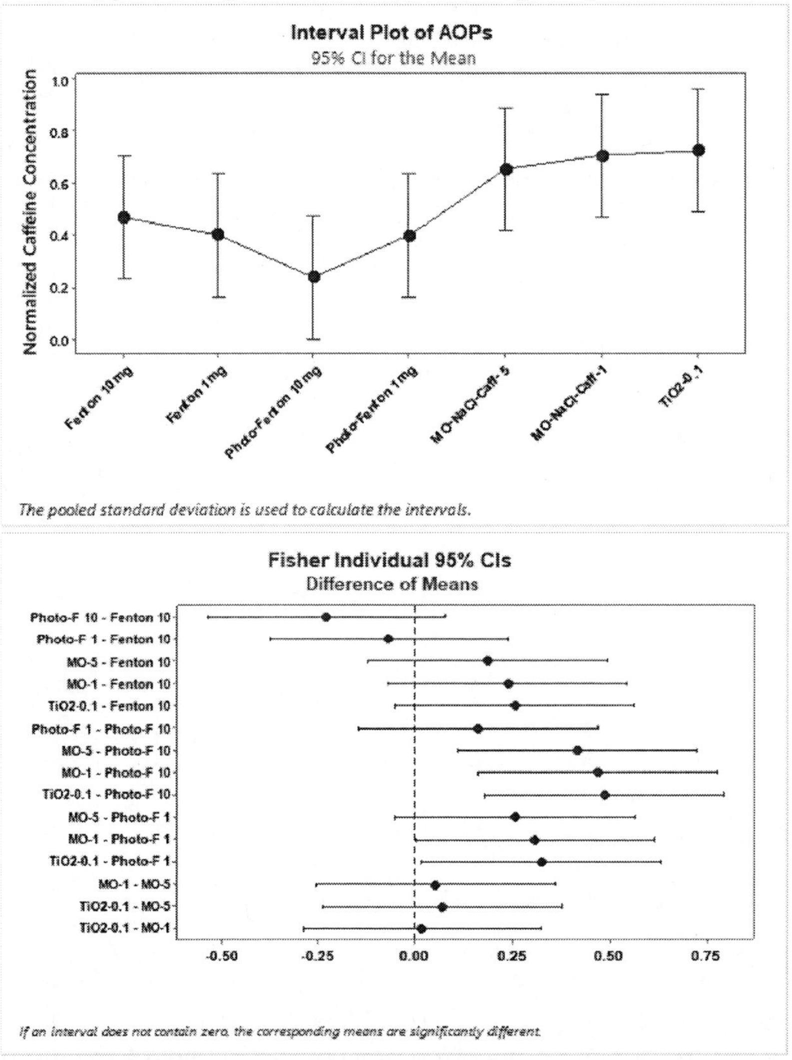

Fig. 2. Statistical results for the caffeine removal by MO-NaCl (1g), Fenton (1mg, 10mg), Photo-Fenton (1mtg, 10mg) (a) ANOVA. 95% CI for the mean of different methods. The most representative methods were found to be significantly different; (b) Fisher's test results. If an interval does not contain zero, the corresponding means are significantly different. Moringa (1 g) and TiO_2 (0.1g) were found to be similar due to their similar caffeine removal.

The different methods were found to be statistically different ($p<0.05$) (Fig. 2(a)), and a Fisher test was carried out (Fig. 2(b)) for the means, which were found to be different. No significant statistical differences were found between TiO_2 and Moringa (1 g of seed extract), which correlates with the fact they both led to similar caffeine removal percentages. However, this comparison is only from a mathematical stand-point, and does not imply any kinetic correlations between procedures, since each of them involve very different Physical-Chemistry processes.

To provide further insight on the difference between the methods under study, we used UV-Vis Through this kind of spectra, differences between AOPs and MO methods are clearly observed. UV spectra showed that as proposed, MO is an adsorption process through flocculation, while Fenton-like processes are based in the degradation of the organic compound as observed in Fig. 3.

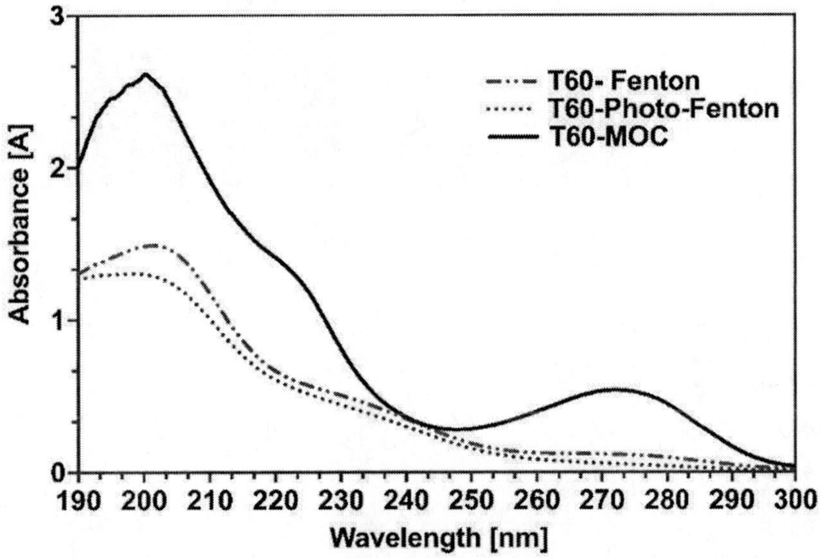

Fig. 3. UV spectra of MO and Fenton-like samples. Caffeine removal with MO whole seed was achieved through a coagulation/flocculation/adsorption process, whereas caffeine removal in Fenton (1 mg, 0.036 mmol/L) and Photo-Fenton (1 mg, 0.035 mmol/L) was achieved through a degradation *process* as evidenced in the intensity decrease of the UV absorption.

3.2 MOC2-NaCl and 5-50 ppm initial caffeine concentrations

The mechanism of caffeine degradation was demonstrated using AOPs, while the removal mechanism using moringa protein extracts was verified as one of coagulation/flocculation, however, moringa extracts were applied to much higher initial concentrations of caffeine (100 ppm) during set-one experiments which is not realistic for wastewater conditions in Quito. For this reason, lower ranges of concentrations were chosen and specifically, with the focus in the 5-ppm initial concentration which is a characteristic value in one of the most important wastewater discharges of the Quito's Metropolitan District area.

The evolution of caffeine fractions remaining in solution as a function of time for the tested MOC2-NaCl dosages is observed in Fig. 4. Also, the final removals after 60 min are evidenced. Initial concentration of artificial caffeine water was 5 ppm.

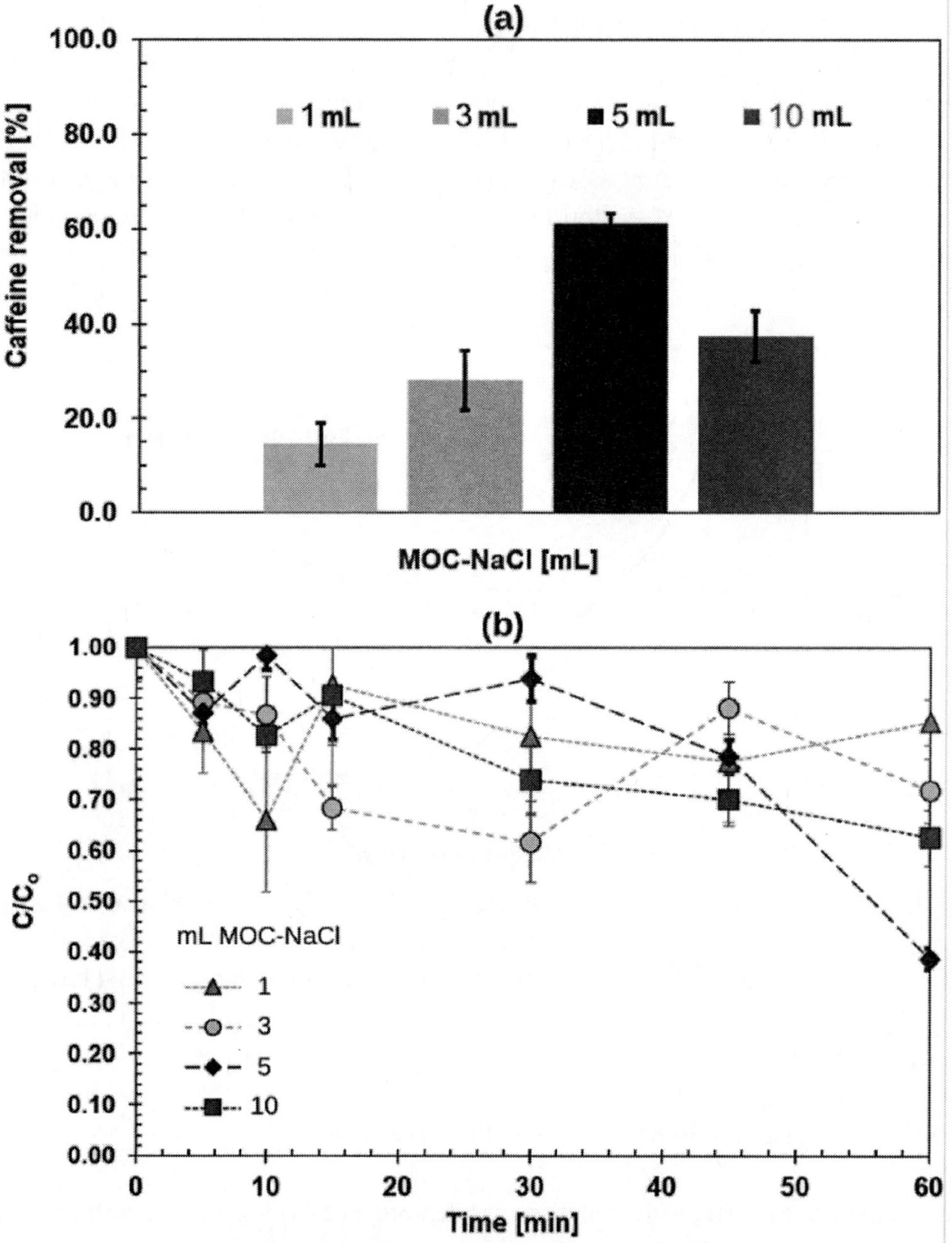

Fig. 4. a) Evolution of caffeine present in coagulation/flocculation experiments as a function of time for all four MOC2-NaCl treatments, b) percent removal after 60 min.

After 60 min, the best dosage with MOC2-NaCl corresponded to 5 mL, resulting in a removal of 61.3 ± 2.1 %. It was found that the removal percentage after 10 min, the 3-mL and 5-mL treatments are statistically the same after 10 min of process. The same occurs for the 30-min case.

For the 60-min time data, the dosages are significantly different, except for the 3-mL and 10-mL cases. This could mark an interval, in which the best removal would be achieved between 3 mL and 10 mL of extract. Likewise, in this time interval, the 5-mL dosage showed better performance, giving an idea of the possible volume in which there will be greater removal with MOC2-NaCl protein extract.

The effectiveness of MOC2-NaCl has been reported for the case of carmine indigo dye in a processing plant, although initial concentrations of pollutant ranged from 40 to 400 mg/L [7]. The study by Beltrán-Heredia, Sánchez-Martín, & Delgado-Regalado (2009), showed up to 80% reduction of indigo carmine dye, which is a blue dye used in the food industry [35], after a 20-min process. This study used 10 mL of a stock solution made of 1000 mg of dye per liter of water along 5 mL of MO seed extract which was then filled up to a 100-mL volume. The initial concentration used in the present work contemplates much lower concentrations, characteristic of caffeine presence in the main wastewater discharges of Quito. Nonetheless, this study was a main reference to test Ecuadorian MOC2-NaCl extracts in caffeine as an emerging contaminant model. Coagulation-flocculation experiments with the same variety of moringa cultivated in Ecuador were performed in a previous study over real water sources samples [10], however, this study used crush MOC of different diameters as a more practical method for wastewater treatment in a single step to be incorporated *in situ* and a reduction of COD, turbidity, and *E. Coli* was assessed already reporting promising results in real water. For instance, the method herein described could be an alternative for targeting emerging contaminants as well. An important aspect to consider, is that residual MO from cold-pressed processes for seed oil extraction is usually available as waste and this residual moringa seed powder (RMSP) could be used as raw material for MOC2-NaCl preparation and water usage could be minimized.

3.3 Degradation with a modified Fenton Process

Fenton reactions were performed according to a modified Fenton Process where the recommended pH value was 2.6 ± 0.2 and 25 °C to favor the solubility of iron, increase the speed of the peroxide reaction and the rate of transformation of the contaminant [28], [30], [36] and since the system has a significant degradation effect due to the oxidation potential of hydroxyl radicals, which is approximately E^o = 1.8-2.7 V [37]. The results of caffeine degradation through Fenton process in artificial water are presented in Fig. 5.

Fig. 5 shows the evolution of the remaining fractions in solution for initial concentrations of 5, 25 and 50 ppm respectively. The Fenton reaction occurred within the 5-10 min interval. This is due to the photolysis of hydrogen peroxide in the aqueous phase which produces radicals and can be applied in the destruction of micro contaminants [38], such as caffeine. It can be verified that AOP is a good procedure to remove ECs, and almost the same removal, around 60% achieved with MOC2-NaCl, is achieved in 5 minutes of process. Nonetheless, the use of MO seed extract does not involve the addition of chemicals besides NaCl, no pH control is required (pH values ranged 6.5-7), the procedure can take place at environmental conditions, which is advantageous, especially for rural community use and/or limiting economic circumstances.

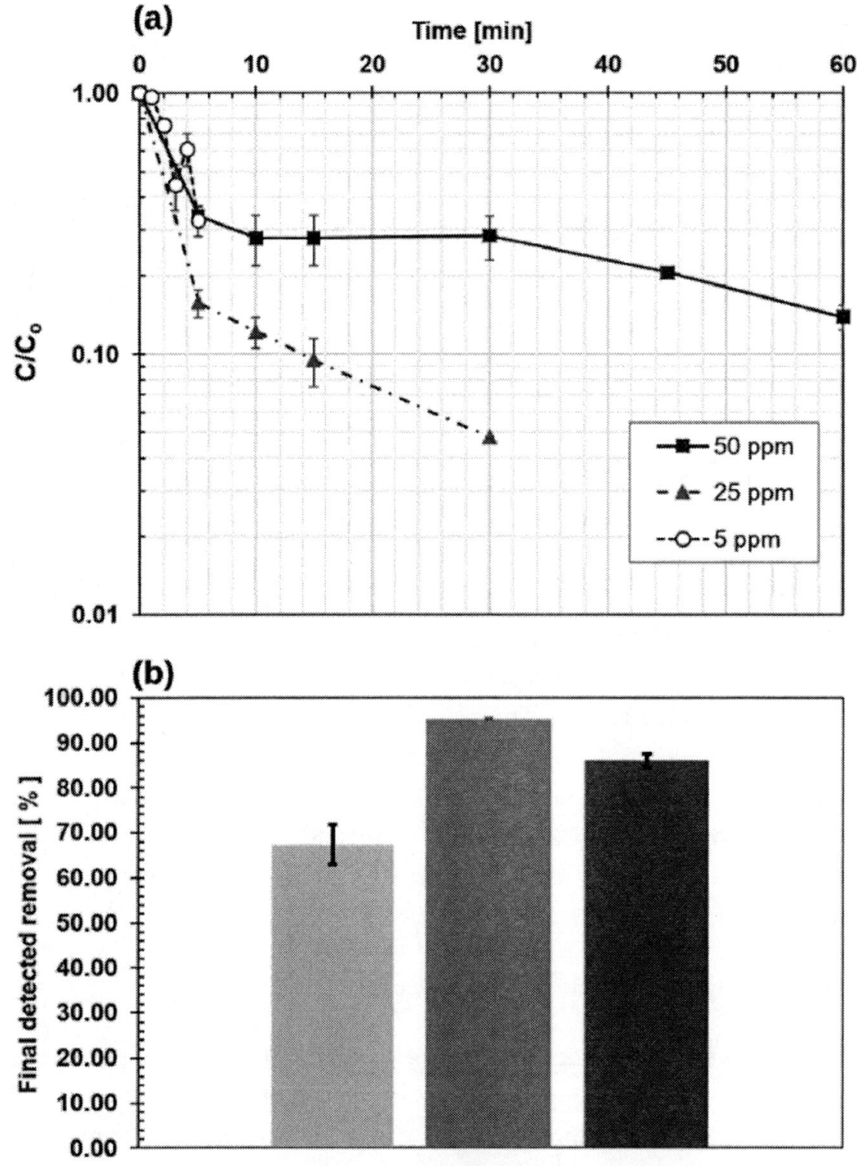

Fig. 5. a) Evolution of caffeine present in coagulation/flocculation experiments as a function of time for 5, 25 and 50 ppm initial concentrations; b) final detected removals after 5 min (5 ppm), 30 min (25 ppm) and 60 min (50 ppm).

As mentioned before, RMSP could also be used for emerging contaminant removal, such as it is suggested in the work by León et al. (2020) were MO residues were impregnated with commercial TiO_2 nanoparticles for irgasan and caffeine removal [39]. Depending on the objective, application and practicality, MO seeds and MOC2-NaCl extracts prove to be a feasible green solution for water treatment applications.

3.4 Dynamic Filtration

Filtration experiments as a function of time and filter-bed depth were performed. A thorough description and discussion of these can be found elsewhere [27] and the most important findings are summarized in what follows. Fig. 6 shows caffeine concentration as a function of time for all filtration media. In general, around 180 s of operation caffeine concentration approaches asymptotically to caffeine's initial concentration of 50 ppm and zeolite present much higher resistance to saturation and the level curves appear to be different for each bed length (L= 13, 18, 23, and 26 cm). In contrast, the level curves for gravel are not significantly different from each other at the same lengths (results not shown), reason why the corresponding Figure for gravel (and moringa husk) were constructed at a wider length range (L= 18 cm -54 cm) to be able to show differences of caffeine concentration as a function of length. In the case of moringa husk, the trend is similar to the other filtration beds, but the error bars are larger. It is hypothesized that this is due to the varying morphology of the husks which provide greater fluctuating results.

In general, caffeine concentration varies linearly with respect to depth up to 27 cm, approximately (Fig. 7). Past this depth, the filtration media loses effectiveness and the rate of change of caffeine as a function of depth decreases as supported by Iwasaki and Ives [40] where the pollutant concentration variation with respect to the bed layer is proportional to its concentration. The concentration gradient is higher at lower operational times. The most distinguishable level curves within all operational times are those of zeolite, and this filtration media leads to the lowest caffeine removal, followed by moringa husk and gravel.

An important aspect to consider during filtration processes, especially when irregular in size materials are used it the relationship of filtration bed mass to bed length; in this way, the amount of caffeine that a specific bed can adsorb can be determined [27]. These ratios were found to be 0.433 [g/cm], 0.873 [g/cm], 0.075 [g/cm], and 0.861 [g/cm] for activated carbon, zeolite, *Moringa oleifera* husk, and gravel respectively. HPLC and UV-VIS results show that caffeine is removed through and adsorption phenomena (Fig. 8) where the characteristic peak of caffeine concentration decreased after treatment, maintaining its original shape.

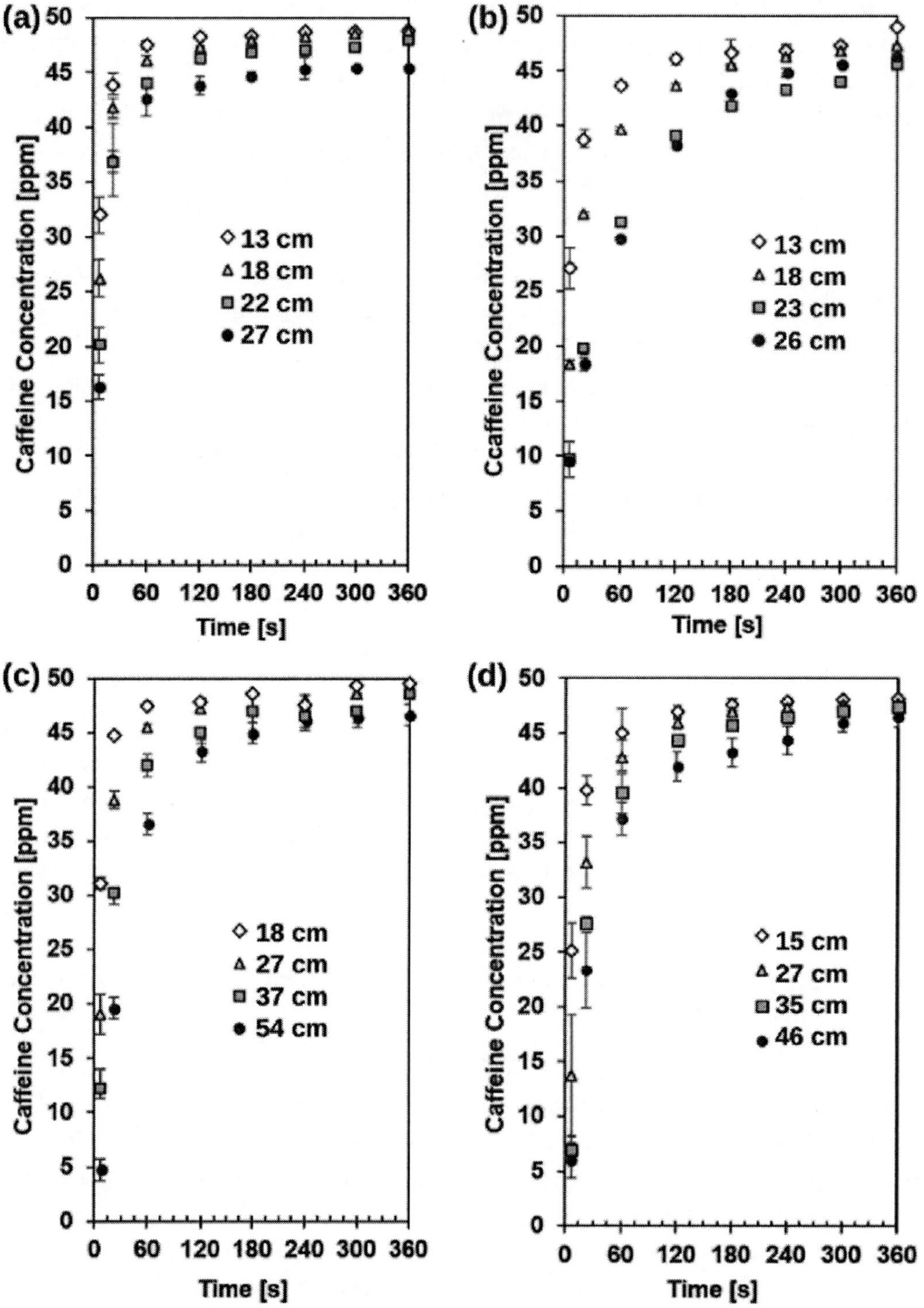

Fig. 6. Caffeine concentration as a function of time for different depth lengths (L) using (a) activated carbon, (b) zeolite, (c) gravel, and (b) moringa.

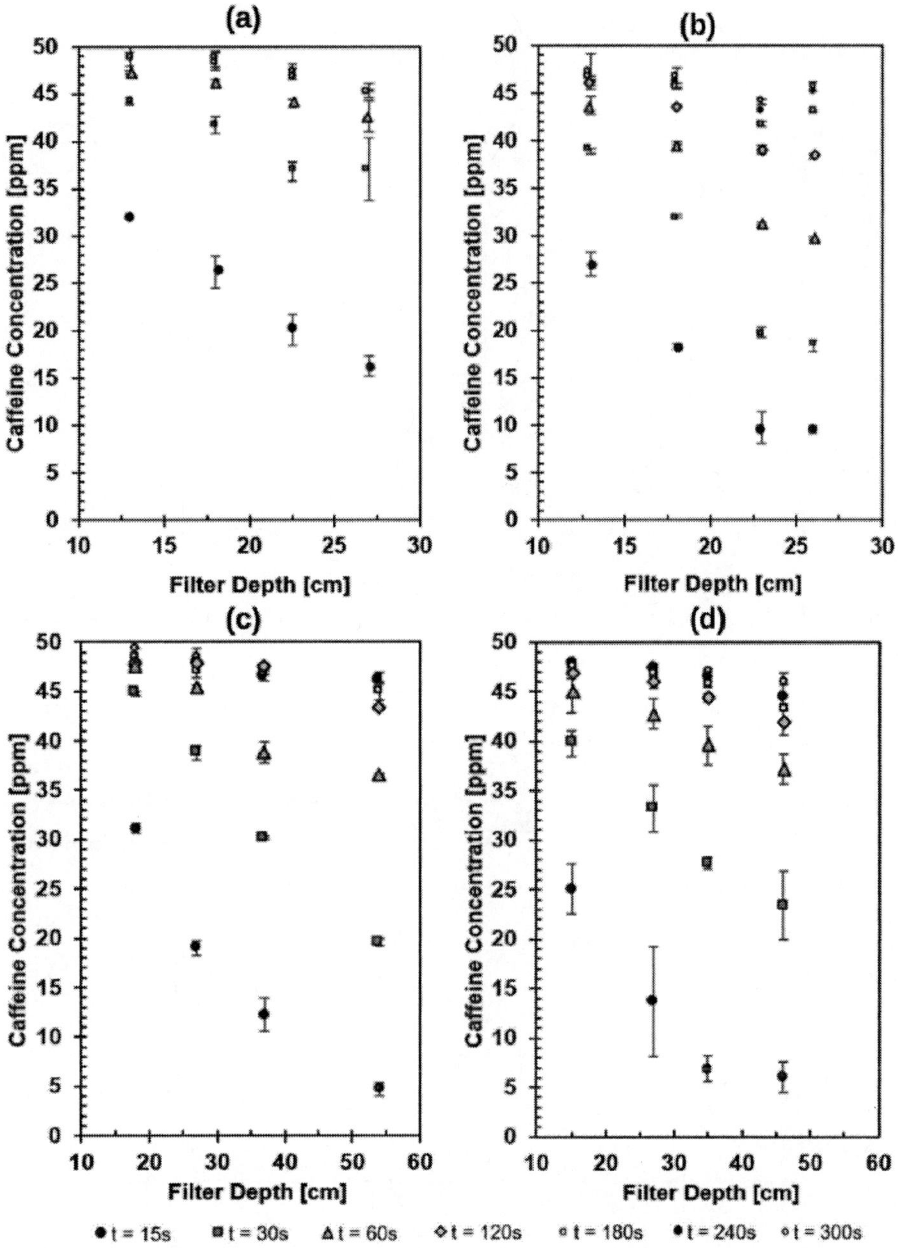

Fig. 7. Caffeine concentration as a function of depth for different operating times (t) using a) activated carbon, (b) zeolite, (c) gravel, and (b) moringa.

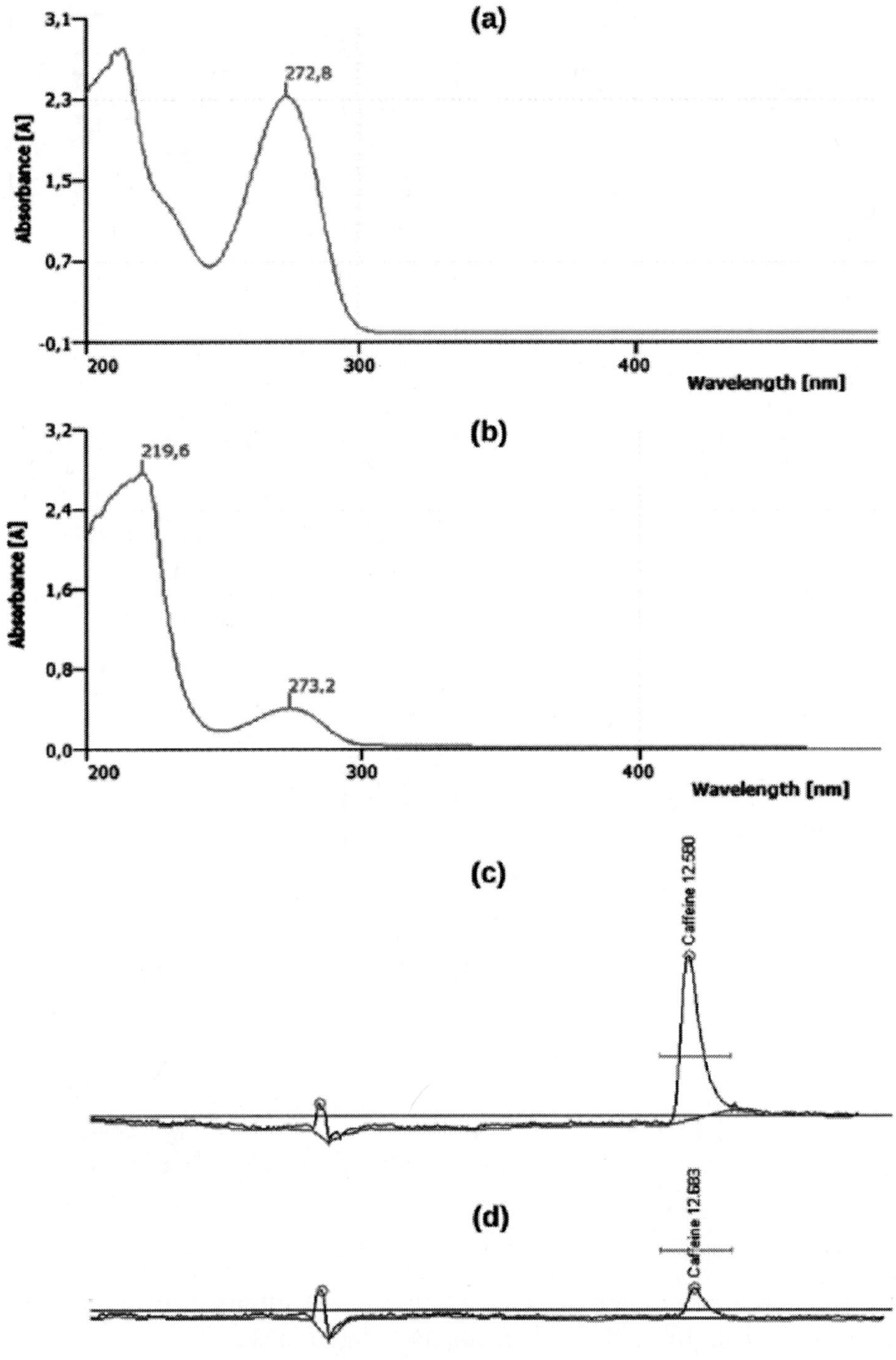

Fig. 8. UV-VIS spectra of a caffeine artificial solution before (a) and after (b) dynamic filtration using Gravel as the filtration medium (t = 10 s, L = 26 cm), HPLC Chromatogram (c) before and (d) after filtration using *Moringa oleifera* Husk as the filtration medium (L = 26 cm).

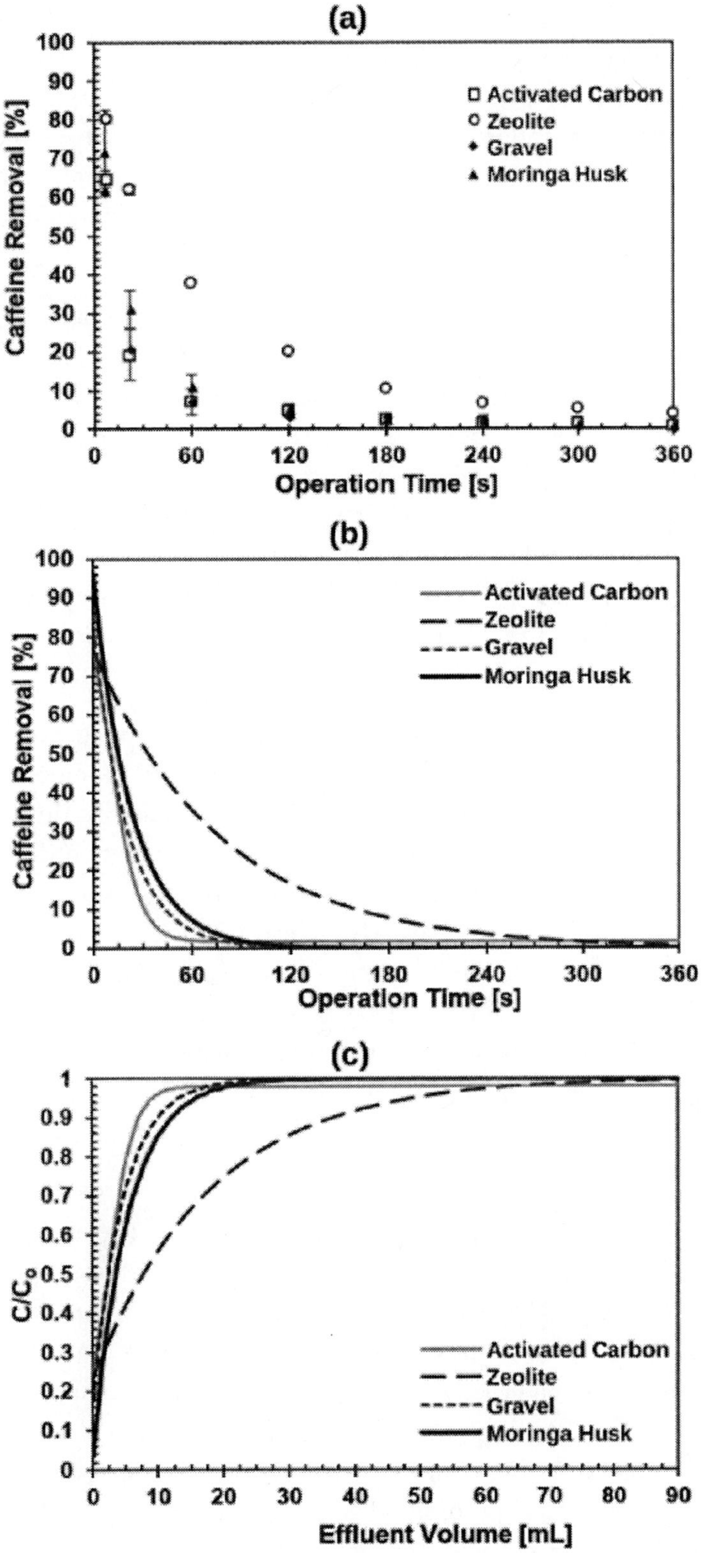

Fig. 9. a) Percentage of caffeine removal as a function of time for L = 27 cm, b) Percentage of caffeine removal as a function of time for L = 27 cm, c) Breakthrough Curves for L = 27 cm.

Next, the previously mentioned filtration models were adjusted to experimental data. The most remarkable conclusions are that the breakthrough curves are best represented by the BBFM except for activated carbon which is represented by Clark's Model. Clark's model assumes that the pollutants follow a Freundlich adsorption isotherm which suggests that a multilayer is formed and technically the contaminants are going to be continuously adsorbed into the bed even past saturation and a larger filter bed implies a larger surface area where caffeine could continuously form a multilayer [27]

As a specific example, considering the 27-cm bed data, the initial caffeine removal is 75%, 81% and 96% for zeolite, gravel and moringa, respectively. Also, from exhaustion point calculations, on average, the mass of caffeine retained per unit bed was calculated to be: 1.100 ± 0.20 (activated carbon), 0.564 ± 0.10 (zeolite), 1.681 ± 0.22 (gravel), and 1.196 ± 0.46 (moringa husk); which is relevant information for possible scale-up purposes. Moringa oleifera husk or gravel will be the preferred materials for instantaneous removals, while zeolite the best material that gives higher removals through time. Fig. 9 shows caffeine removal as a function of time and breakthrough curves when the best fitted models are applied (e.g. for $L = 27$ cm). The complete set of results for all depths can be found elsewhere [27].

Similar trends in shape have been found in other studies involving dyes or heavy metals but with higher adsorption capacities [7], [41], [42], as well as more evident breaking points which occur at lower flowrates [15]. It is worth noticing that in this work the materials were used as they were, without any other pretreatment since practical and cheap solutions are emphasized. Evidence shows that further treatments of the materials are necessary to gain higher efficiencies Also, packing efficiency is expected to increase by reducing particle size and increasing packing compression. Varying the superficial area of the beds as a function of particle size and hydraulic loss calculations will complement these findings and could help determine which combination yields the highest caffeine removal for each filtration media.

3.5 Sustainability

Since 1987, when Brundtland et al. defined sustainability as the fact that human kind should take care of the environment and the resources "meeting the needs of the present without compromising the ability of future generations to meet their own needs" the global challenge has been to find ways towards a sustainable manufacturing through green processes to produce goods and commodities for humans assuring a clean environment. "Sustainable" is nothing else than maximizing economics . and social aspects, minimizing environmental impacts, replacing fossil resources with renewable feedstocks, as well as closing cycles [43]. Natural water cycle has been profoundly altered by human activities causing a deterioration in water quality in pristine waters through urban and industrial discharges into pristine water bodies [44]. This situation along with the increasing demand of clean water due the growing population and industrial expansion force to redesign and rethink sustainable wastewater treatment methods to recover emergent pollutants [45]. Sustainable consumption and production refer to the use of products and services in such a way that respond the basic needs and enhance the quality of life while using minimum natural resources, wastes, as well as toxics and pollutants, over the life cycle of the product or service [43]. Therefore the use of lignocellulosic wastes from agroindustries as adsorbent materials to reduce water contaminants is committed to sustainability as it accomplishes two purposes: it reduces the lignocellulosic waste negative impact in the environment and allows and efficient and sustainable use of a natural renewable resource, which otherwise does not enter in the value chain of the agroindustry production; and in the other hand minimizes the emergent pollutants in contaminated water, making it safer for consumption. In this way, being focused on sustainability, carbon foot print may be reduced and communities may live in healthier living conditions.

4. Conclusions

This study focuses on the application of advanced oxidation, filtration and coagulation/flocculation processes using moringa seed wastes for caffeine removal, as a model emerging organic contaminant. The concentrations of interest ranged 5-100 ppm. A 5-ppm concentration reflects the approximate caffeine content in one of the most important wastewater discharges in the Metropolitan District of Quito, the same that contributes around 33% of the city's total wastewater discharges. A dosage between 3-10 mL of MOC-NaCl is recommended to naturally achieve 61.28 ± 0.021% (5 mL) removal after one hour of process. Caffeine degradation by Fenton reaction occurs mainly within the first 5 to 10 min (67.35 ± 0.043%). The optimal pH of AOPs should be in a range of 2.6 ± 0.2, since if it is less than the value within the standard deviation, the efficiency of the reaction decreases. However, the process described is technically a viable and green alternative to remove ECs.

Different filtration models were tested and compared to experimental dynamic data, where caffeine solutions at 50 ppm initial concentration were used. The Clark model presented the best fit for the case of activated carbon while the BBFM model adapted better for zeolite, gravel and moringa husk. Zeolite is the appropriate candidate if the objective is to remove the maximum amount of caffeine, while moringa husk will be the best option if the most caffeine is to the be retained with the least amount of bed mass.

Acknowledgement

This project was funded by the USFQ Collaboration Grant 2017 and Poli-Grants 2020-2021 and 2021-2022.

References

[1] C. Von Sonntag, "Advanced oxidation processes: Mechanistic aspects," *Water Science and Technology*, vol. 58, no. 5, pp. 1015–1021, 2008, doi: 10.2166/wst.2008.467.

[2] E. E. Ebrahiem, M. N. Al-Maghrabi, and A. R. Mobarki, "Removal of organic pollutants from industrial wastewater by applying photo-Fenton oxidation technology," *Arabian Journal of Chemistry*, 2013, doi: 10.1016/j.arabjc.2013.06.012.

[3] T. D. de Oliveira, W. S. Martini, M. D. R. Santos, M. A. C. Matos, and L. L. da Rocha, "Caffeine Oxidation in Water by Fenton and Fenton-Like Processes: Effects of Inorganic Anions and Ecotoxicological Evaluation on Aquatic Organisms," *J Braz Chem Soc*, 2014, doi: 10.5935/0103-5053.20140237.

[4] Z. Song, N. Wang, L. Zhu, A. Huang, X. Zhao, and H. Tang, "Efficient oxidative degradation of triclosan by using an enhanced Fenton-like process," *Chemical Engineering Journal*, vol. 198–199, pp. 379–387, 2012, doi: 10.1016/j.cej.2012.05.067.

[5] G. Iervolino, I. Zammit, V. Vaiano, and L. Rizzo, "Limitations and Prospects for Wastewater Treatment by UV and Visible-Light-Active Heterogeneous Photocatalysis: A Critical Review," *Top Curr Chem*, vol. 378, no. 1, p. 7, 2019, doi: 10.1007/s41061-019-0272-1.

[6] A. D. Bokare and W. Choi, *Review of iron-free Fenton-like systems for activating H2O2in advanced oxidation processes*, vol. 275. Elsevier B.V., 2014. doi: 10.1016/j.jhazmat.2014.04.054.

[7] J. Beltrán-Heredia, J. Sánchez-Martín, A. Delgado-Regalado, and A. Beltran-Heredia, J., Sánchez, J., & Delgado, "Removal of Carmine Indigo Dye with Moringa oleifera Seed Extract," *Ind Eng Chem Res*, vol. 48, no. 14, pp. 6512–6520, 2009, doi: 10.1021/ie9004833.

[8] A. F. S. Santos *et al.*, "Removal of tetracycline from contaminated water by *Moringa oleifera* seed preparations," *Environ Technol*, vol. 3330, no. December, pp. 1–8, 2015, doi: 10.1080/09593330.2015.1080309.

[9] K. A. Ghebremichael, K. R. Gunaratna, H. Henriksson, H. Brumer, and G. Dalhammar, "A simple purification and activity assay of the coagulant protein from Moringa oleifera seed.," *Water Res*, vol. 39, no. 11, pp. 2338–44, Jun. 2005, doi: 10.1016/j.watres.2005.04.012.

[10] A. C. Landázuri *et al.*, "Experimental evaluation of crushed Moringa oleifera Lam. seeds and powder waste during coagulation-flocculation processes," *J Environ Chem Eng*, vol. 6, no. 4, pp. 5443–5451, Aug. 2018, doi: 10.1016/J.JECE.2018.08.021.

[11] G. S. Madrona, M. R. S. Scapim, L. A. C. Tonon, M. H. M. Reis, C. M. Paraiso, and R. Bergamasco, "Use of Moringa oleifera in a combined coagulation-filtration process for water treatment," *Chem Eng Trans*, vol. 57, no. 2011, 2017, doi: 10.3303/CET1757200.

[12] C. N. Malusare and R. M. Gidde, "Study of Moringa Oleifera Extracts in Water Treatment," *National seminar vision 2025, technological developments in biological sciences*, pp. 1–12, 2011.

[13] K. Shebek, A. B. Schantz, I. Sines, K. Lauser, S. Velegol, and M. Kumar, "The Flocculating Cationic Polypetide from Moringa oleifera Seeds Damages Bacterial Cell Membranes by Causing Membrane Fusion," *Langmuir*, vol. 31, no. 15, pp. 4496–4502, Apr. 2015, doi: 10.1021/acs.langmuir.5b00015.

[14] N. B. G. Salguero, "Characterization of contaminants by different treatments for their removal in water : Use of Caffeine as a model compound," Universidad San Francisco de Quito, 2016.

[15] J. L. Sotelo, A. Rodríguez, S. Álvarez, and J. García, "Removal of caffeine and diclofenac on activated carbon in fixed bed column," *Chemical Engineering Research and Design*, vol. 90, no. 7, pp. 967–974, 2012, doi: 10.1016/j.cherd.2011.10.012.

[16] T. P. Izod, "REMOVAL OF CAFFEINE BY SELECTIVE ADSORPTION USING ZEOLITE ADSORBENTS," 1982.

[17] M. Bruni and D. Spuhler, *Slow Sand Filtration*. 2012. Accessed: Feb. 08, 2022. [Online]. Available: https://sswm.info/water-nutrient-cycle/water-purification/hardwares/semi-centralised-drinking-water-treatments/slow-sand-filtration

[18] Tech Brief, "Slow Sand Filtration," 2000.

[19] R. Bird, W. Stewart, and E. Lightfoot, *Transport Phenomena*. John Wiley & Sons. Inc., 2002.

[20] H. C. Thomas, "Heterogeneous Ion Exchange in a Flowing System," *J Am Chem Soc*, vol. 66, no. 10, pp. 1664–1666, Oct. 1944, doi: 10.1021/ja01238a017.

[21] R. Ansari, B. Seyghali, A. Mohammad-Khah, and M. A. Zanjanchi, "Highly efficient adsorption of anionic dyes from aqueous solutions using sawdust modified by cationic surfactant of cetyltrimethylammonium bromide," *J Surfactants Deterg*, vol. 15, no. 5, pp. 557–565, Sep. 2012, doi: 10.1007/s11743-012-1334-3.

[22] R. Han *et al.*, "Adsorption of methylene blue by phoenix tree leaf powder in a fixed-bed column : experiments and prediction of breakthrough curves fixed-bed column : experiments and prediction," no. September 2015, 2009, doi: 10.1016/j.desal.200.

[23] J. Peric, M. Trgo, and N. Vukojevic, "Testing of Breakthrough Curves for Removal of Lead Ions from Aqueous Solutions by Natural Zeolite-Clinoptilolite According to the Clark Kinetic Equation," no. October 2007, pp. 944–959, 2008, doi: 10.1080/01496390701870622.

[24] A. C. Landázuri, J. S. Villarreal, J. C. Andrade, I. Sornoza, and A. S. Lagos, "Bulk balance filtration model (BBFM) for lead and iron physisorption through Moringa oleifera Lam. seed husks," *J Environ Chem Eng*, vol. 7, no. 5, p. 103302, 2019, doi: 10.1016/j.jece.2019.103302.

[25] EPMAPS, "Obras de Intercepción y Tratamiento de las Aguas Residuales para Quito y sus Parroquias," 2016.

[26] a. Voloshenko-Rossin *et al.*, "Emerging pollutants in the Esmeraldas watershed in Ecuador: discharge and attenuation of emerging organic pollutants along the San Pedro–Guayllabamba–Esmeraldas rivers," *Environ. Sci.: Processes Impacts*, vol. 17, no. 1, pp. 41–53, 2015, doi: 10.1039/C4EM00394B.

[27] I. Sornoza Merino, "Caffeine Removal through the use of a Dynamic Filter with Moringa oleifera Lam. and common filtering beds," Universidad San Francisco de Quito, Quito, 2017. Accessed: May 06, 2022. [Online]. Available: https://repositorio.usfq.edu.ec/bitstream/23000/7113/1/135735.pdf

[28] M. R. Rojas, F. Pérez, D. Whitley, R. G. Arnold, and A. E. Sáez, "Modeling of advanced oxidation of trace organic contaminants by hydrogen peroxide photolysis and fentons reaction," *Ind Eng Chem Res*, vol. 49, no. 22, pp. 11331–11343, 2010, doi: 10.1021/ie101592p.

[29] R. Khan, *Problem solving and data analysis using minitab: A clear and easy guide to six sigma methodology.*, 1st. United Kingdom: Wiley, 2013.

[30] J. J. Pignatello, E. Oliveros, and A. MacKay, "Advanced oxidation processes for organic contaminant destruction based on the fenton reaction and related chemistry," *Crit Rev Environ Sci Technol*, vol. 36, no. 1, pp. 1–84, 2006, doi: 10.1080/10643380500326564.

[31] A. G. Trovó *et al.*, "Degradation of caffeine by photo-Fenton process: optimization of treatment conditions using experimental design.," *Chemosphere*, vol. 90, no. 2, pp. 170–5, Jan. 2013, doi: 10.1016/j.chemosphere.2012.06.022.

[32] E. Yamal-Turbay, M. Graells, and M. P??rez-Moya, "Systematic assessment of the influence of hydrogen peroxide dosage on caffeine degradation by the photo-fenton process," *Ind Eng Chem Res*, vol. 51, no. 13, pp. 4770–4778, 2012, doi: 10.1021/ie202256k.

[33] R. R. N. Marques *et al.*, "Photocatalytic degradation of caffeine: Developing solutions for emerging pollutants," *Catal Today*, vol. 209, pp. 108–115, 2013, doi: 10.1016/j.cattod.2012.10.008.

[34] H. A. Jerri, K. J. Adolfsen, L. R. McCullough, D. Velegol, and S. B. Velegol, "Antimicrobial sand via adsorption of cationic Moringa oleifera protein," *Langmuir*, vol. 28, no. 4, pp. 2262–2268, 2012, doi: 10.1021/la2038262.

[35] M. de Keijzer, M. R. van Bommel, R. H. Keijzer, R. Knaller, and E. Oberhumer, "Indigo carmine: Understanding a problematic blue dye," *Studies in Conservation*, vol. 57, no. sup1, pp. S87–S95, 2012, doi: 10.1179/2047058412Y.0000000058.

[36] E. Kan and S. G. Huling, "Effects of temperature and acidic pre-treatment on Fenton-driven oxidation of MTBE-spent granular activated carbon," *Environ Sci Technol*, vol. 43, no. 5, pp. 1493–1499, 2009, doi: 10.1021/es802360f.

[37] J. A. Hangasky, T. C. Detomasi, C. M. Lemon, and M. A. Marletta, "5.12 - Glycosidic Bond Oxidation: The Structure, Function, and Mechanism of Polysaccharide Monooxygenases," H.-W. (Ben) Liu and T. P. B. T.-C. N. P. I. I. I. Begley, Eds. Oxford: Elsevier, 2020, pp. 298–331. doi: https://doi.org/10.1016/B978-0-12-409547-2.14859-0.

[38] T. A. Tuhkanen, "UV/H2O2 Preocesses. In Advanced Oxidation Processes for Water and Wastewater Treatment.," *Parsons, S.*, 2004.

[39] C. León, G., Aldás, M., Guerrero, V., Landázuri, A., & Almeida-Naranjo, "Caffeine and irgasan removal from water using bamboo, laurel and moringa residues impregnated with commercial TiO2 nanoparticles," *MRS Adv*, vol. 4, no. 64, pp. 3553–3567, 2019, doi: 10.1557/adv.2020.33.

[40] A. Valencia, *Teoría, Diseño y Control de los Procesos de Clarificación del Agua*. Lima, Perú: CEPIS (Centro Panamericano de Ingenieria Sanitaria y Ciencias del Ambiente), 1972.

[41] J. S. Villarreal, J. R. Gándara, D. Navarrete, M. L. Bejarano, and A. C. Landázuri, "Lead (Pb2+) adsorption by means of pristine and prewashed residual Moringa oleifera Lam. seed husk biomass for water treatment applications," *International Journal of Sustainable Engineering*, vol. 00, no. 00, pp. 1–13, 2020, doi: 10.1080/19397038.2020.1862350.

[42] A. C. Landázuri, J. Cahuasqui, and A. Lagos, "Metal adsorption in aqueous media using Moringa oleifera Lam. seeds produced in Ecuador as an alternative method for water treatment," *Avances en Ciencias e Ingenierías, USFQ*, vol. 11, no. 2, pp. 190–205, 2019, doi: http://dx.doi.org/10.18272/aci.v11i2.

[43] U. N. E. Programme, *Sustainable Production and Consumption - A Handbook for Policymakers*. 2015.

[44] L. Sala and M. Serra, "Towards Sustainability in Water Recycling," *Science and Technology*, vol. 50, no. 2 (November), pp. 1–7, 2004, doi: 10.2166/wst.2004.0074.

[45] S. W. Hermanowicz, "Sustainability in water resources management: Changes in meaning and perception," *Sustain Sci*, vol. 3, no. 2, pp. 181–188, 2008, doi: 10.1007/s11625-008-0055-z.

393a Integrated Sustainability Assessment: Exergy, Emergy, Life Cycle Assessment

Natalia Cano, Camilo A. Franco, Farid B Cortés, Krisztián Baracza, Jo Dewulf, Heriberto Cabezas

The word sustainable is a controversial concept because it has been defined in different ways by many disciplines. However, as of the 1980s, the word is used to refer to an appropriate management of natural resources, in such a way as to allow future generations the access to adequate resources. This definition was articulated in the Brundtland Report (Brundtland commission, 1987). In the Bruntland Report, the idea that a sustainability analysis interrelating economic development and environment was developed. Later, at 2005 World Summit on Social Development, objectives of sustainable development were identified: economic development, social development and environmental protection, also known as the three pillars of sustainability (United Nations General Assembly, 2005). Nearly all governments have in principle committed themselves to sustainable development by integrating economic welfare, environmental quality, and social coherence (Böhringer, C.; Jochem, 2007). The preservation of the natural environment is a prerequisite for a well-functioning economy and social justice. Therefore, sustainable development means something more than a compromise between the natural environment and pursuit of economic growth (Reza, Sadiq and Hewage, 2014). It means a definition of development which recognizes that sustainability limits have a structural origin both (economic) and natural.

Different methodological tools for sustainability assessment have been observed. Some oriented only to environmental sustainability based not only on the environmental burden (generated in the Life Cycle Assesment production chain) but also on the use of direct and indirect resources provided by nature for its manufacture, ending in the conception of the final product, and the investment and economic return that provides social welfare (Emergy Analysis, EmA). Others that include process efficiency in terms of entropy generated (Exergy Analysis, ExA). All these methodologies converge to the same point: tools that provide to the decision-makers environmental, economic or social sustainability indicators for the formulation and implementation of public policies. Another meeting point of these analytical methods is that each one has exactly the same methodological limitations: joint production, commensuration, deliberation, boundaries system setting, and double counting. However, this methodological tools that account for sustainability indicators in environmental, economic and / or social dimensions, cannot provide an assessment under these three dimensions in a robust way by themselves. Since some of them develop indicators that others do not, it is necessary to implement them as a complement but not as interchangeable techniques, providing additional information for decision making.

In recent years, there has been an increasing tendency to 'integrate' different accounting methods to have a more complete picture of impacts caused by the creation of a product or service provision, instead of depending on a method that generally has a single criterion perspective (Kharrazi *et al.*, 2014). These indicators can be taken as individual or composite parameters; that is, synthetic aggregations of independent parameters, reflecting the stakeholders values or considerations (Arbault *et al.*, 2014). Regardless of whether individual or composite indicators are taken, a sustainable process should present an energy and technical feasibility that generates economic gains for social welfare with an acceptable environmental burden, hence the need to make use of all tools for assessing this sustainability, not as exchangeable but as complementary. A holistic view of the supply and value chain sustainability is necessary to ensure adequate management through an integrated approachable to assess processes from two points of view: 'user-side' and 'donor-side'. User-side analyzes final efficiency indicators of the process (environmental impacts per unit of product generated, or energy /

exergy delivered per unit of energy /exergy inverted). Donor-side considers the work nature does to provide resources for the production of a good or service, and in turn, to assimilate the pollutant load released to the environment as an important component of sustainability assessment. Accounting of the economic return in relation to the investment made cannot be ignored; as well as the effects and welfare that project execution can bring to society (Arbault *et al.*, 2014).

Detractors of this integration argue that it is a reductionist approach, which leads to an assessment of the three sustainability dimensions through non-correlated factors (Arbault *et al.*, 2014; Kharrazi *et al.*, 2014). Although a holistic approach may result in the loss of information about each individual dimension and the indicators that compose them, it is also true that creating a holistic system-level image of the interactions between dimensions is critical for quantifying sustainability. The challenge is then to build methodological tools that can be integrated quantitatively into three dimensions.

Research aim is to propose a sustainability assessment framework to obtain a unified performance metric (Integrated Sustainability Index, ISI) to assess Triple Bottom Line – TBL. To achieve this integration, Life Cycle Assessment (LCA) and Exergy Analysis (ExA) as User-side approach, and Emergy Accounting (EmA) as donor-side view are implemented, in spite of the fact that each method differs on its basis of analysis and solve common methodological limitations in a different way. Some of the main issues faced by both designers and decision makers are precision and uncertainty in the calculation of these integrated indexes. This research presents systematically the proposed methodology to achieve the desired clarity and transparency. Methodological calculations supporting can be found at (Cano, 2018; Cano Londoño, Velásquez and McIntyre, 2019a, 2019b). The proposed sustainability index presents a comprehensible hierarchic structure determined by support methodologies such as LCA, ExA, and EmA, which have regulatory and academic validity. The integration of social, environmental, and economic components in an index that also allows the subjective adjustment of externalities reducing the risk of subjectivity, overshadowing rigorous work, is a new approach to assessing sustainability in development projects

393b Interaction Analysis in Multiscale Sustainability Assessment

Mahboubeh Moghadasi, Yinlun Huang

Sustainability assessment is a key step towards achievement of industrial sustainability. Over the past two decades, numerous types of sustainability metrics and assessment methods have been developed, most of which are for the performance assessment of products, processes, plants, and beyond; from the multiscale systems science and engineering point of view, the assessment is at the macroscopic level. It is recognized that some types of sustainability problems associated with products and processes may be rooted at the meso- or even microscopic levels. For instance, some novel products using nanomaterials have been identified with various environmental and health problems and economic and social challenges. Those problems and challenges could and should be addressed during the product design stage at the much finer scales of length and time. Take nanopaint design and nanocoating manufacturing as an example, the selection, shape, orientation, and volume fraction of nanoparticles, the selection of polymer material, solvent and additives, and the interfacial phenomena between nanoparticles and polymer materials during nanomanufacturing are all related to the eventual sustainability performance at the macroscopic level. This renders a need for extending the sustainability assessment to the meso and microscopic levels.

Major challenges in multiscale sustainability assessment include: (i) how to identify interactions among indicators in different sustainability categories at different scales, (ii) whether there is a need and how to unify sustainability indicators at different scales, (iii) how to determine a minimal set of indicators at the different scales, and (iv) how to conduct an effective multiscale sustainability assessment. In this paper, we introduce a framework for analyzing systematically interactions among multiscale sustainability indicators, determining a minimal set of multiscale sustainability metrics, and conducting deep sustainability assessment by resorting sustainability engineering principles. The framework enables elimination of redundancy, while ensuring comprehensiveness of sustainability performance analysis. A case study on the multiscale sustainability assessment of nanopaint design and nanocoating manufacturing will be presented to demonstrate methodological attractiveness.

393c Optimization Based Selection of Industrial Biosolids Treatment and Valorization Pathways to Meet International Discharge Standards for Heavy Metals

Nivinya Hemachandra, Dhabia Al-Mohannadi, Debalina Sengupta

Biosolids can be classified into residential or domestic sludge and industrial sludge. While residential sludge undergoes various reuse pathways, industrial bio-sludge typically ends up in landfill sites due to their heavy metal contamination. However, there exists multiple ways such as chemical extraction of heavy metals, absorption, microwave induced heavy metal removal, adsorption, and biosorption for treatment. At the same time, biosolids can be converted biochar, incinerated for energy production, or used in agriculture land application if it meets regulatory limits. These processes vary in cost, scale, efficiency of the process, energy, emissions, material, and energy demand. This work presents an optimization-based approach that screens different type of technologies for treatment and reuse of bio-sludge/biosolids with the objective of minimizing costs while adhering to the regulatory limits from various countries. A case study on industrial Gas-to-Liquid (GTL) waste biosolids was used to validate the proposed approach. The model identifies the cost optimal pathway of removing the heavy metal content off industrial biosolids to meet the ceiling concentration required. The biosolids then pass-through general treatment pathways which returns the biosolid contents as value added products which can be used as fertilizer or sold to other entities.

393d Modeling and Analysis of Anaerobic Membrane Bioreactor (AnMBR) for Integrated Resource Recovery from Wastewater

Madison E. Kratzer, Prathap Parameswaran, Vikas Khanna

Concepts such as the Water-Energy Nexus and Circular Economy are becoming increasingly important in global resource management. To achieve these goals integrated resource recovery from wastewater is a high priority. Anaerobic Membrane Bioreactor (AnMBR) is an emerging technology that allows for the digestion of waste and capture the products of digestion including energy, nutrients, and highly value-added chemicals. AnMBRs require the use of costly membranes that need to be regularly replaced, and the costs of membrane cleaning and replacement may limit the industrial application of this technology. Captured products of digestion may allow for the recuperation of operating costs and allow AnMBR to be an economically viable alternative to traditional wastewater treatment techniques. To evaluate the viability of AnMBR technology a detailed process model must be developed. Additionally, due to the variability in microbiological communities and membrane fouling dynamics of AnMBR systems, results of AnMBR treatment must be evaluated for specific waste streams.

We present a robust process model based on the ADM1 model for the treatment of swine wastewater in an AnMBR system. Updates to the ADM1 model have been included to incorporate membrane fouling dynamics and potential inhibition of methanogenesis through the consumption of acetate. The process model has been developed so that optimization of operating conditions can be conducted at steady-state equilibrium conditions of the digester. Specifically optimization using mixed integer nonlinear programing (MINLP) is conducted in Pyomo using the IPOPT algorithm. Sensitivity analyses show the importance of membrane fouling, solids retention time, and initial concentration of carbohydrates, lipids, and proteins in the waste stream. The implications of the findings for recovering resources from animal wastewater streams will be discussed along with discussion of future studies.

393e A Closed Loop Case Study of Food Waste Management in Singapore

Yen Wah Tong, Hailin Tian

Food waste (FW) shows an increasing trend worldwide in recent years. If not treated properly, it causes other problems such as water and air pollutions. In Singapore, FW is an outstanding issue due to the large population residing in a limited island and the considerable yearly increasing FW quantity with low recycling rates (<20%). Singapore has a unique food culture that more than 100 food courts are distributed in the island for residents' daily meal consumption (Tong et al., 2018). Hence, food waste onsite treatment to save the transportation cost and emissions has become a consensus. Anaerobic digestion (AD), as a widely used technology to recover energy and resources from different waste, has been used to treat food waste in large scale in many countries. However, small-scale decentralized AD system in urban area to specifically treat food waste onsite is seldom reported. A comparative life cycle assessment study performed by Tian et al. (2021) showed the environmental benefits of deploying decentralized system, but the actual operational performance and cost analysis are yet to be further explored. Moreover, in most of the food waste treatment studies, the organic loading is normally fixed during certain experimental periods. As for the actual situation of a food court, it is difficult to ensure the same daily organic loading, particularly the different waste quantity during weekdays and weekends, thus a robustness testing of a decentralized system fed with real and varying food waste quantity from a food court needs to be validated. Therefore, the aim of this study is to 1) customize a decentralized AD system to treat food waste for a representative food court in Singapore, i.e., East Coast Lagoon Food Village (ECLFV), and 2) discuss the robustness, energy balance, carbon emission profile, and cost benefits, of the system.

The electricity consumption by the system and the electricity generation by the biogas engine were recorded. The operation results showed that the electricity consumption that is not related to the organic loading but keeps the system running was about 9kwh per day, including circulation pump, control system, ventilation fans, etc., while the electricity consumption that is related closely to the organic loading was about 0.09kwh/kg FW, including the feeding pump, blender, digestate discharge pump, biogas compressor, etc. The engine converts the biogas into electricity at a rate of 1.78m3 biogas/kwh (STP), i.e., the electricity efficiency was around 10.5%. The efficiency is lower compared to other studies, which might be due to the small scale of the engine. At this stage with dine-in restriction, the low FW collection resulted in low electricity generation, which the recovered energy only contributed 12% of the system consumption. With the release of the restriction, the FW collection is expected to be close to the waste audit results. Hence, based on the waste audit results and the current system performance, the recovered energy is estimated to reach at around 16kwh/day, while the electricity consumption is also estimated to reach at 28kwh due to the high FW collection and feeding, which the recovered energy contribution can increase from 12% to 59%. Therefore, to have the system as electricity self-sustaining system, more effects should be spent on the improvement of the biogas engine efficiency to at least 18%.

312a Bioproducts from High-Strength Wastewater with the Carboxylate Platform Via Methane-Arrested Anaerobic Digestion

Haoran Wu, Meltem Urgun-Demirtas, Mark Holtzapple, Phuong Thai, Rachel Dalke, Lauren Valentino

In circular economy, valorizing low- and negative-value waste streams into valuable products are essential to minimize waste and pollution, reduce dependency on fossil fuels, and combat climate change. With the growth of global economy, large quantities of high-strength wastewater (HSW) are discharged into the environment, resulting in severe problems. When treating HSW, traditional treatment processes are hindered by low treatment efficiency or high operation costs. Thus, a new cost-effective HSW treatment process needs to be developed to efficiently valorize HSW with low carbon footprint.

The carboxylate platform incorporating methane-arrested anaerobic digestion (MAAD) is a promising alternative to sustainably produce high-value bio-derived chemicals and fuels from waste streams. In this study, a novel process based on the carboxylate platform was developed to simultaneously treat HSW with bio-based carboxylic acid production.

This study firstly addresses technical issues with process development with a focus on MAAD. Specifically, a robust microbial consortium with high salt toxicity tolerance was selectively established and enriched to achieve high waste conversion and acid productivity. A blend stream of dairy and brewery wastewater with elevated chemical oxygen demand (COD) concentration (>70 g/L) was tested as the feed HSW. The effects of wastewater characteristics, microbial consortia structure, digester operation mode (batch and semi-continuous), and operating conditions (e.g., temperature, substrate concentration, and pH) on MAAD performance were investigated at bench scale (500-mL). The MAAD process was then scaled up and optimized at lab-scale digester (14-L), and the most stable condition of hydraulic residence time (HRT) 3 days, pH 6.0, and 40 °C was selected for further scale-up. Semi-continuous pilot-scale MAAD (100-gal) produced a total acid concentration of around 40 g/L with COD conversion greater than 70%. A new kinetic model was developed to predict the total acid productivity in semi-continuous MAADs. Different cost-effective separation methods (e.g., membrane, adsorption, electrodeionization) were evaluated to determine the most suitable approach to harvest carboxylic acids from MAAD broth. Further, this study compares different downstream technology approaches with the aid of techno-economic assessment (TEA). The results show that the newly developed process has the potential for large-scale application and is a successful example of the waste-to-energy technologies to transform low- or negative-value waste streams into high-value bioproducts.

312c Engineering a New Self-Regenerative System to Achieve Effective Removal of Heavy Metal Contaminants from Polluted Waters

Sweta Roy, Zehui Han, Dacheng Ren

About 77 million Americans are exposed to contaminated water that contains heavy-metal toxins such as arsenic, lead and mercury. Ranked first on the Superfund List of Hazardous Substances (CDC) is arsenic (As). Arsenic is an ubiquitous toxic metalloid that can replace substitution of phosphates and interact with protein thiol groups; thus, interrupting major biochemical pathways and cleaves DNA leading to high risk of skin, kidney, lung and bladder cancers. Currently, millions of people are exposed to toxic levels of As, predominately through drinking water. Conventional techniques used for As removal from contaminated waters are limited by high cost, secondary pollutants, incomplete removal, and lack of metal specificity. Therefore, an effective method for removal is necessary. Here we engineered a new method of remediation by magnetizing arsenic for removal (M.A.R). Specifically, we have engineered a bacterial strain with a surface-displayed peptide, AfArsR, which binds specifically to arsenite As(III). In addition, the bacterium was engineered with a capability to respond to magnetism, allowing for an easy removal of As(III) ions from the water. Expression of the surface display was validated using a fluorescent protein and the engineered cells exhibited good magnetic response. The findings from this study will facilitate the design of new treatment facilities for economical remediation of water contamination as well as unwanted toxins in the environment.

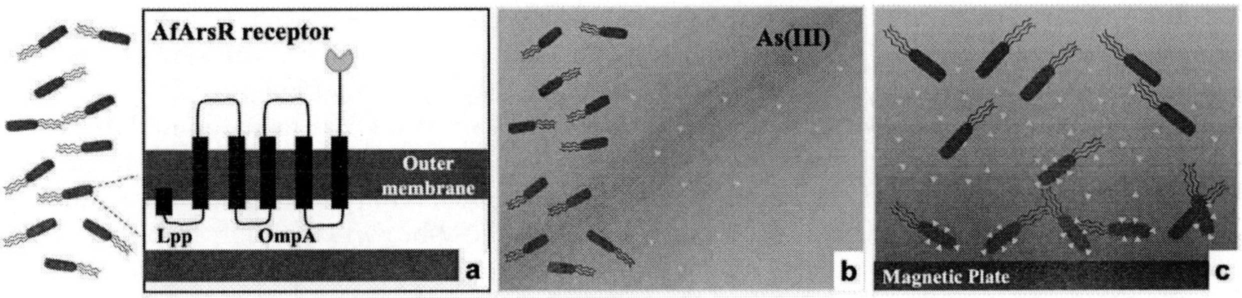

312d Lipase Immobilization on Mesoporous Carbon and Evaluation As Catalysts in the Hydrolysis of Waste Oleochemical Streams

Alexander Baena, Alvaro Orjuela

The increasing global generation of used cooking oils (UCOs) is resulting in a cascade of economic, environmental, and public health issues as mismanagement is a common practice worldwide. Hence industrial exploitation and valorization is required towards a circular economy within the oleochemical industry. However main challenges for UCOs harnessing are their heterogeneity and the high content of impurities, acidity, and water. Alternatively, waste UCOs could be subjected to complete hydrolysis to recover free fatty acids (FFAs), which are widely used oleochemical feedstocks. Different catalysts have been explored for this process; however, lipases are of particular interest because they are able to work on low-quality feedstocks under milder operating conditions and with high selectivity, thus reducing capital and energy costs and the associated environmental impacts. On the other hand, the high pressure and temperature hydrolysis of refined vegetable oils and fats is typically used in the manufacture of fatty acids, this process is highly energy-intensive, and the severe conditions can drive a variety of undesirable side reactions. Despite the high conversion, selectivity, and benign conditions, the industrial implementation of the enzymatic processes is still challenging owing to the high cost of the enzymes, the lower reaction rates compared with homogeneous catalysts, and the deactivation that can be exacerbated by impurities contained in the feedstock.

In order to improve the enzymatic processes, immobilization techniques have been used to enhance enzymes stability and selectivity, simultaneously facilitating recovery and reutilization. As the immobilized enzymes are more resilient to reactive media, they are then effectively used as catalysts in traditional and ultrasound-assisted reactors [1] [2]. At the same time, the use of immobilized catalysts can help to overcome the stability issues when processing waste streams as the feedstock of the hydrolysis process. This is particularly important because reclaiming and exploiting waste oleochemical streams (e.g. used cooking oil, trap greases, yellow-brown-black fat, crop mill effluents) is paramount to reducing major environmental impacts on the food/feed production chain [3]. Currently, there are commercial immobilized enzymes, but they are characterized by low yields and slow kinetics in the hydrolysis of refined oils and fats [4]. This occurs partly because the enzymes have been immobilized under microporous and highly hydrophobic materials (e.g. silica, alumina, zeolites).

In this regard, the immobilization of enzymes on mesoporous and hydrophilic supports seems a feasible pathway to enhance their effectiveness in the hydrolysis of waste oleochemical streams. First, three different immobilization method were tested, showing that amino-functionalized activated carbons have an enormous potential for either further lipase immobilization or activation with glutaraldehyde. The biocatalyst obtained were tested in hydrolysis reactions, for the experiments ultrasound-assisted conditions at 37 and 80 kHz at 37 ºC were used, since reaction media is a critical point different molar ratios of water to oil (9:1, 5:1, 1:1 and 1:3) and lipase concentration were studied. On the other hand, it is well known that long exposure time might induce denaturation for this reason experiments with and without ultrasound and mechanical stirring were carried out. For comparison purposes, the reaction was carried out under the best hydrolysis conditions previously reported for a commercial lipase (Novozyme® 435). It was found that the immobilized lipases on the highly hydrophobic mesoporous material had better performance on the hydrolysis of waste oleochemicals than the commercial one, in terms of yield and reusability, in the same way high contents of water may allow to obtain hydrolysis degrees up to 90% but with low reusability, this indicates that hydrophilic medium induces enzyme leaching, it was also identified that ultrasound has no influence in the structural changes in the lipase during the reaction.

312f Fed-Batch Screening of Methanotroph-Algae Cocultures on Anaerobic Digester Effluent for Larger-Scale Wastewater Treatment and Valorization

Loyal Murphy, Kiumars Badr, Riley O'Gwynn, Q. Peter He, Jin Wang

Anaerobic digestion (AD) is a mature waste-to-value technology and is commonly installed in large-scale water resource recovery facilities across the United States. The utilization of AD produced biogas is limited to heat and electricity generation due to contaminants such as hydrogen sulfide, ammonia, and volatile organic compounds. In addition, the eutrophic AD effluent, rich in nitrogen and phosphorous, has to be further treated before releasing back into the environment.

Recently, it has been demonstrated that methanotroph-microalgae cocultures offer a highly effective platform for biogas valorization and nutrient recovery from wastewater [1-3]. As different microorganisms usually exhibit a wide range of native capabilities in terms of growth and tolerance to inhibitors contained in AD effluent, it is important to select and optimize the coculture species for the development of successful waste-to-value biotechnologies. Specifically, the coculture pairs should demonstrate fast and robust growth in the presence of different inhibitors and competitors contained in the AD effluent without sterilization.

To facilitate the selection of the proper coculture strains for the integrated biogas valorization and nutrient recovery, we have developed a screening station which contains nine parallel apparatuses that enable simultaneous testing of multiple species (**Figure 1**). As temperature, pH, gas feed composition and agitation all have significant effect on the growth of microbes, the screening station must provide nearly identical environments for consistent screening. In addition, light intensity must be regulated for the screening of microalgae and coculture pairs. To achieve uniform conditions across all nine apparatus, different control elements were installed. The variations of temperature and agitation rates were controlled within ±1.0K and ± 10rpm through temperature-controlled stir plates (Four E's Scientific). The composition and flowrate of inlet gas were controlled via an in-house gas mixing system [4] and mechanic flowmeters (Cole-Parmer, Masterflex® Correlated Variable-Area Flowmeters). The variation in pH was controlled within ±0.1 through BlueLab pH controller. Lastly, an LED light wall with reflecting mirrors was constructed to provide an even amount of light to each vessel to facilitate photoautotrophic growth. Light intensity was quantified using a PAR meter (Quantum) to verify that each vessel was receiving roughly the same photon flux density within ±10% of the average.

AD effluent from the mesophilic digestor at Columbus Water Works (Columbus, GA) was collected on October 12, 2021, and frozen until needed. Seven methanotrophs and seven photoautotrophs (**Table 1**) that have been studied for wastewater treatment were selected for screening. Triplicate fed-batch screenings of individual microalgal and methanotrophic species on diluted wastewater revealed which species had the best candidacy for the coculture. Factors taken into consideration were maximum growth rate and organic carbon production. The top three candidates from each species were tested as cocultures (nine cocultures in all) to determine a robust coculture for wastewater remediation. Coculture experiments were also taken in triplicate.

Each 500mL vessel contained 250mL of working volume, starting with an initial optical density of roughly 0.1. Individual vessels were constructed to provide airtight, sterile environments for cell growth. Experiments for this testing proceeded until each culture was at least one day into the stationary phase. The OD was recorded several times a day, while organic carbon measurements were only taken at the start and end of experiments.

References:

[1] Roberts N, Hilliard M, He QP and Wang J (2020) *Front. Energy Res. 8*, 563352.

[2] Li X, Lu Y, Li N, Wang Y, Yu R, Zhu G and Zeng RJ. (2022) *Env. Sci. andamp; Tech. 56* (6), 3669-

3677.

[3] Ruiz-Ruiz P, Gomez-Borraz TL, Revah S and Morales M (2020) *Chemosphere. 259*, 127418.

[4] Stone, K.A., He, Q.P. andamp; Wang, J. (2019) *Sci. Rep.* 9, 5899.

Table 1: The species of methanotrophs and photoautotrophs studied in this work

Methanotrophs	Photoautotrophs
Methylococcus Capsulatus	*Chlorella Sorokiniana*
Methylosinus Trichosporium OB3b	*Chlorella Zofingiensis*
Methylomonas sp. LW13	Chlorella Vulgaris
Methylosarcina Fibrata	*Chlorella Kessleri*
Methylosarcina Quisquiliarum	*Scenedesmus Obliquus*
Methylosarcina Lacus sp. nov. LW14T	*Scenedesmus Dimorphus*
Methylomonas Methanica S1	*Scenedesmus sp.*

Figure 1: Image of the nine reaction apparatuses used for experimentation. Red arrows and subtitles are provided for added clarity

312g Recycling PVC Containing Wood Wastes for Value-Added Products

Xianglan Bai, Xiaolin Chen

Polyvinyl chloride (PVC)-containing waste streams are difficult to recycle due to a high chlorine content in PVC. Toxic dioxin or corrosive hydrogen chloride vapor released from improper management of PVC-containing wastes can cause severe environmental pollution and human health problems. While PVC is usually treated as contamination and burden in waste recycling, a novel recycling approach was developed in this study to leverage PVC as an asset. Specifically, red oak and PVC were co-converted in γ-Valerolactone, a green biomass-derivable solvent. During the co-conversion, PVC-derived hydrogen chloride in the solvent acted as an acid catalyst to produce up to 14.4% levoglucosenone and 14.3% furfural from red oak. On the other hand, dechlorinated PVC hydrocarbons and lignin fraction of red oak reacted to each other to form chlorine-free solid fuels with high thermal stability. The higher heating value of the solids was up to 36.2 MJ/kg, which is even higher than the heating value of anthracite coal. After the co-conversion, more than 80% of PVC-contained chlorine turned into chloride ion and the rest formed 5-chlorovaleric acid. 5-chlorovaleric acid crystals were further recovered from the post-reaction liquid with a purity of 91.2%.

312h One-Pot Conversion of PVC Containing Co-Mingled Wastes

Xianglan Bai, Xiaolin Chen

Co-mingled wastes containing organic and synthetic polymer materials are difficult to recycle due to their complex compositions. Each year, billion tons of co-mingled wastes are sent to landfills each year, causing significant environmental and economic burdens. Recycling polyvinyl chloride (PVC) containing wastes is particularly challenging due to the release of toxic chlorinated dioxins and corrosive acid when recycled. In this study, cellulose, PVC and non-PVC plastics such as PE, PP and PS, were co-converted in tetrahydrofuran (THF). During the co-conversion, PVC-released HCl acted as an in-situ acid catalyst to promote cellulose conversion to levoglucosan. It showed that the presence of non-PVC plastic polymers facilitates the cellulose conversion by PVC-derived acid through influencing the solvent acidity. After co-conversion, dechlorinated solids containing non-PVC polymers and cellulose-derived char are recovered. The highest HHV of 46.34 MJ/kg was obtained in the solid produced the PP containing wastes, which are even better than HHV of diesel (44.4 MJ/kg). SEM analysis shows hydrocarbons and cellulose-derived char were chemically interweaved into porous solids. When the solids were used as oil absorbents, they demonstrated much superior absorbent capacities than hydrochars derived from the same waste feedstocks.

445a Techno-Economic Optimization of Electrochemical Reactors for CO_2 Reduction

Lucas Cammann, Isabell Bagemihl, Maria Del Mar Pérez-Fortes, Volkert van Steijn, J Ruud Van Ommen

The production of base chemicals, such as various hydrocarbons, by electrochemical conversion of CO_2 paired with CO_2 capture technologies has the potential to close the carbon cycle, thereby significantly contributing to the future energy transition. The feasibility of electrochemical CO_2 conversion has been demonstrated at lab scale, leading researchers to optimize reactor designs based on performance targets derived from techno-economic analysis. However, these techno-economic analysis are commonly based on fixed performance metrics and therefore neglect the well studied interdependence of crucial performance parameters such as current density, faradaic efficiency and cell potential on the mechanistic level. This confines the techno-economic studies to univariate sensitivity analyses assuming that reactor performance parameters are independent of each other, thereby overestimating the solution space for feasible reactor designs.

This study, therefore, presents a multi-scale modelling approach spanning from a firstprinciple reactor model over the electrolyser-stack scale to the process scale, allowing us to assess the well described interdependencies on the reactor design level from an economic perspective. The first-principle model of a CO_2 electrolyser is developed based on experimental data in literature for the production of ethylene and captures the interplay between CO_2 conversion, consumption and faradaic efficiency for varying current densities and inlet velocities. This framework does not only show the propagation of interdependencies across scales it further enables us to perform the first techno-economic optimization for CO_2-electrolysis systems. The optimal performance targets are found to strongly deviate from previously reported targets, which manifests for the herein chosen reactor design, in an optimal current density of $\approx 100 mAcm^{-2}$ in comparison to commonly reported values greater than $200 mAcm^{-2}$. This approach highlights the importance of including interdependencies on the reactor scale to asses the economic viability of a reactor design and presents a tool to evaluate reactor design choices. This study further shows that the accuracy of these predictions is highly dependent on the available data, which makes it important to *a)* move towards more holistic,multi-scale modelling approaches in the field of CO_2-electrolysis, and to *b)* communicate measured or targeted reactor performance with all applicable boundary conditions.

445b A Continuous, Integrated Electrochemical System for CO_2 Capture and Valorization

Rohan Sartape, Aditya Prajapati, Meenesh Singh

The heavy reliance on energies derived from fossil fuels has led to an increase in atmospheric CO_2 concentration air by over 40% compared to the pre-industrial era. The steady increment of CO_2 has led to a heavy contribution to global warming and disturbance to flora and fauna. Therefore, it is urgent to endorse a carbon-neutral process. Integrated CO_2 capture and production of value-added products is an attractive approach for leveraging monetary benefits from a carbon-neutral process. Herein, we report the development of a continuous, integrated electrochemical system for capturing CO_2 and valorization to produce economically significant fuels and products. In an Integrated CO_2 Capture and Valorization (ICCV) process, the CO_2 capture occurs in a migration-assisted moisture gradient (MAMG) unit where CO_2 is bound with an organic liquid and transported to an aqueous reservoir in the presence of an electric field. The CO_2 valorization process occurs in an electrochemical CO_2 reduction unit that utilizes a copper mesh catalyst and aqueous reservoir from the CO_2 capture unit to produce value-added fuels and products such as H_2, CH_4, CO, C_2H_4. Herein, we report the effect of impurities in flue gas such as SO_x and NO_x, operating current densities for electrochemical CO_2 capture and electrochemical reduction on the ICCV system. Since these parameters significantly impact the faradaic efficiencies of the electrochemical reduction reaction, a complete product distribution from the long-term operation of ICCV is studied using HPLC and GC.

445c Process Design and Optimization of Absorption-Based CO_2 Capture Processes with a Low-Pressure Flash Column

Haryn Park, Joohwa Lee, Seokwon Yun, Jin-Kuk KIM

With tightening restrictions for climate change and growing implementation efforts for carbon-neutrality, the importance of the hydrogen economy has considerably increased. However, SMR(Steam Methane Reforming) based technology using natural gas, which is mostly used for hydrogen production, has an environmental limitation to achieve the net zero carbon emissions. To realize the sustainable production of hydrogen at industrial scale, it is essential to introduce CO_2 capture process on SMR-based hydrogen plants.

This study aims to design and simulate the amine-based CO_2 capture process for the reduction of energy consumption when CO_2 capture is applied for a high-pressure syngas stream generated from a shift converter reactor. The new design strategies are considered, including adding a low-pressure flash column and introducing a semi-lean amine stream. Optimization and sensitivity analysis by changing the key design variables are carried out. Assessment for technical feasibility on the configurational changes is conducted, which evaluates the techno-economic benefit of the proposed configuration against the traditional absorber-stripper arrangement.

Acknowledgement

This work was supported by the National Research Foundation of Korea (NRF) grant funded by the Korea government (MSIT) (No. 2019R1A2C2002263) and by the Korea Institute of Energy Technology Evaluation and Planning (KETEP) and the Ministry of Trade, Industry andamp; Energy (MOTIE) of the Republic of Korea (No. 20214710100060).

445f Multi-Objective Optimization of Carbon Capture Utilization and Storage Supply Chain

Manar Oqbi, Dhabia Al-Mohannadi

The world is challenged to reduce greenhouse effect which is a main contributor to global warming. Greenhouse gases (GHG) emissions are increasing due to human activities. Accordingly, there is a need to reduce GHG emissions and limit global climate change. Carbon capture utilization and storage (CCUS) became increasingly essential to reduce carbon dioxide emissions from different sectors around the world. There is a challenge to design sustainable CCUS that considers safety, economics, and environmental aspects of the CCUS supply chain. This work presents a methodology for a systematic design of the carbon capture utilization and storage supply chain considering the trade-off between different scales including economics, safety, and environmental emissions. This work is the first to propose a method for quantifying hazards considering different utilization options using safety indices. The multi-objective optimization utilizes augmented Ɛ-constraint method to generate Pareto optimal solutions. The framework is demonstrated by applying it to a CCUS supply chain case study with different sources and sinks to minimize the total annual cost, environmental emissions as well as hazards. The generated Pareto optimal surface provides several unique designs of the CCUS. The proposed approach will help decision makers to select the sustainable optimal design of the supply chain considering safety, economic and environmental impacts of the CCUS.

445g Rate-Based Dynamic Modeling and Analysis of an Amine-Based Carbon Capture Unit for Flexible Operation

Paul Akula, John C. Eslick, Debangsu Bhattacharyya, David Miller

Achieving net-zero carbon emissions is one of the greatest challenges for human society. Modeling and simulation of advanced multiscale process units have great potential in accelerating the commercialization of carbon capture technologies[1]. Amine-based CO_2 capture is the baseline technology for retrofitting existing power stations. However, the integration of amine-based post-combustion CO_2 capture technology with power plants to reduce greenhouse gas emissions incurs a high energy penalty decreasing a power plant's efficiency by about 23 percentage points[2]. Understanding the capture plant dynamics plays an important role in its technical and economic performance. Capital and operating costs of capture plants strongly depend on the configuration[3]. For a given configuration, operating conditions play a vital role in reducing the operating costs[4]. Flowrate and composition of the flue gas change as the host power plant ramps up and down to meet required power output. Hence, various strategies can be considered for improving the economics of the capture unit. For example, loaded solvent can be stored when the price of electricity is high and regenerated when the price drops. Thus, the capture unit must be operated optimally under dynamic, off-design conditions. For dynamic optimization, an accurate dynamic model that captures off-design operation of the amine-scrubbing unit is required. Rate-based dynamic models of amine scrubbing units are limited in the open literature compared with steady-state models. Most existing dynamic rate-based models have been developed assuming linear driving force across the liquid and gas films[5-8]. This assumption can lead to considerable inaccuracy under off-design conditions due to nonlinearity in the transport variables in the film region, especially under off-design dynamic operation. Furthermore, existing rate-based dynamic models consider simple thermodynamic models and constant heat of reaction that do not adequately represent the complex thermodynamics of electrolyte systems.

In this study, a dynamic model of a packed tower is developed for a monoethanolamine(MEA)-based capture unit considering simultaneous mass transfer and chemical reactions in the films. Multi-component transport of molecular and ionic species is modeled using an extended Maxwell-Stefan (MS) transport equation that includes the effect of electrostatic forces for reactive absorption process. A comprehensive description of the thermodynamic framework for electrolyte systems represented by the electrolyte-Non-Random Two Liquid is presented where analytical expressions for excess enthalpy is developed that can improve the accuracy of the enthalpy model for these highly nonlinear systems compared to numerical approaches for computing excess enthalpy. The dynamic model is validated using transient data from the National Carbon Capture Center in Alabama, USA. For dynamic optimization, the simultaneous approach is adopted over the sequential approach by fully discretizing in space and time[9]. While the fully discretized approach offers many advantages, including direct Jacobian and Hessian calculation within the optimizer and efficient decomposition strategies that can exploit structure and sparsity of the system of equations[9], this approach leads to a large-scale optimization problem. The problem is solved using the flexible, open-source Institute for the Design of Advanced Energy Systems Integrated Platform (IDAES)[10] which provides access to efficient large-scale NLP solvers. Several approaches are developed for generating good initial guesses for future state and algebraic variables. In addition, capabilities in IDAES for activating and deactivating constraints are exploited to develop a sequential initialization strategy. Our results show that dynamic optimization not only reduced the energy usage, but also reduces mass transfer limitations thus improving the economics of capture processes under fast, part-load operations.

References

1. Miller, D. C.; Siirola, J. D.; Agarwal, D.; Burgard, A. P.; Lee, A.; Eslick, J. C.; Nicholson, B.; Laird, C.; Biegler, L. T.; Bhattacharyya, D., Next Generation Multi-Scale Process Systems Engineering Framework. In *Computer Aided Chemical Engineering*, Elsevier: 2018; Vol. 44, pp 2209-2214.
2. Supekar, S. D.; Skerlos, S. J., Reassessing the efficiency penalty from carbon capture in coal-fired power plants. *Environmental science andamp; technology* **2015,** *49* (20), 12576-12584.
3. Bhattacharyya, D.; Miller, D., Post-combustion CO 2 capture technologies — a review of processes for solvent-based and sorbent-based CO 2 capture. *Current Opinion in Chemical Engineering* **2017,** *17*, 78-92.
4. Akula, P.; Eslick, J.; Bhattacharyya, D.; Miller, D. C., Model Development, Validation, and Optimization of an MEA-Based Post-Combustion CO2 Capture Process under Part-Load and Variable Capture Operations. *Industrial andamp; Engineering Chemistry Research* **2021**.
5. Montañés, R. M.; Flø, N. E.; Nord, L. O., Dynamic process model validation and control of the amine plant at CO2 Technology Centre Mongstad. *Energies* **2017,** *10* (10), 1527.
6. Ziaii, S.; Rochelle, G. T.; Edgar, T. F., Dynamic Modeling to Minimize Energy Use for CO2 Capture in Power Plants by Aqueous Monoethanolamine. *Industrial andamp; Engineering Chemistry Research* **2009,** *48* (13), 6105-6111.
7. Harun, N.; Nittaya, T.; Douglas, P. L.; Croiset, E.; Ricardez-Sandoval, L. A., Dynamic simulation of MEA absorption process for CO 2 capture from power plants. *International Journal of Greenhouse Gas Control* **2012,** *10*, 295-309.
8. Nittaya, T.; Douglas, P. L.; Croiset, E.; Ricardez-Sandoval, L. A., Dynamic modeling and evaluation of an industrial-scale CO2 capture plant using monoethanolamine absorption processes. *Industrial andamp; Engineering Chemistry Research* **2014,** *53* (28), 11411-11426.
9. Biegler, L. T.; Zavala, V. M., Large-scale nonlinear programming using IPOPT: An integrating framework for enterprise-wide dynamic optimization. *Computers andamp; Chemical Engineering* **2009,** *33* (3), 575-582.
10. Institute for the Design of Advanced Energy Systems (IDAES) https://idaes-pse.readthedocs.io/en/stable (accessed 4/1/2021).

445h Global Renewable Energy and Negative Emission Potential Observatory - in a Knowledge Graph Context

Lanyu Li, Xiaonan Wang

Under the common vision of humanity to reduce carbon emissions, renewable energy and negative emission technologies are promising technologies in the blueprint of low-carbon transition of energy systems. However, hybrid renewable energy systems combined with negative emission technologies, which have a synthetic effect and promising potential in building low-carbon energy systems, are still under-explored. In our recent study, we found that sustainable energy systems combining renewable energy and negative emission technologies offer considerable environmental benefits compared to conventional energy systems and are economically competitive in areas with abundant renewable resources. The proposed decision support tools can be readily applied by end-users, operators, and governments, and the findings from the regional-scale analysis will help countries refine their emissions reduction targets, agree on global emissions reduction commitments, and design sustainable energy systems that benefit the world as a whole.

However, practically, the selection, design, and placement of appropriate renewable energy and negative emission technologies within a given place is a complex, multi-domain, and multi-objective problem that necessitates an efficient knowledge management framework. One possible approach to utilize real-time and multi-domain data for this is The World Avatar (TWA) project. TWA is a dynamic knowledge graph (dKG) based on the Semantic Web and its associated technologies, with intelligent agents operating on it. The agents act autonomously to constantly update and extend TWA, thus making it evolves in time. The purpose of this paper is to illustrate how digital twinning (of the energy system, renewable and negative emission resources, and the technology and market conditions) in the context of a dKG can support and augment our previous work to develop a decision-support system that determining the optimal placement of technologies that is economic and carbon-reducing for the given context. The system consists of a bi-objective robust optimization model – to minimize both the cost of deploying the technologies and maximize the carbon reduction potential that the technologies would bring to the area. A case study considering the climate mitigation of renewable energy and negative emission technologies in China and the UK is introduced to demonstrate the application of this approach.

593a Adsorption of Microcystin-LR on Biochar Studied Using Molecular Simulations

Hemant Nagar, Toufiq Reza, Nirupam Aich, Sumit Sharma

Eutrophication of Great Lakes and coastal regions of the United States is responsible for frequent and unpredictable outbursts of harmful algal blooms (HABs). During HABs, uncontrolled growth of cyanobacteria (blue-green algae) occurs, which releases a toxin called Microcystin-LR in water. Even small concentrations of Microcystin-LR (> 1 μg/l) in the water supply make it unfit for human consumption. A way of reducing eutrophication is to recover nutrients from biowaste by converting them into a nutrient-rich hydrochar via Hydrothermal Carbonization (HTC). Furthermore, pyrolyzed hydrochar is a highly porous and hydrophobic material that can act as an adsorbent of nutrients and Microcystin-LR from water streams. The goal of this research is to employ molecular dynamics (MD) simulations to understand the adsorption behavior of Microcystin-LR on hydrochar surfaces of different chemistries. Our simulation system comprises of microcystin-LR, water, ions and biochar surfaces described in full-atomistic detail. Using advanced simulation methods like umbrella sampling, we are determining the adsorption free energy of Microcystin-LR on the hydrochar surfaces. By employing a new free energy sampling methodology, developed in our research group, we are determining the equilibrium adsorption morphologies of Microcystin-LR on hydrochar surfaces. We are also investigating the aggregation behavior of Microcystin-LR molecules in the bulk aqueous phase. Results from our simulations will complement the adsorption studies that are being performed by our experimental collaborators.

593b Thermochemical Conversion in Space: Developing Mars Regolith-Activated Hydrochar Composite for Enhanced Water Retention in Mars

Al Ibtida Sultana, Toufiq Reza

While the increasing global population translates an increasing need of freshwater on Earth, researchers have been recently applying hydrothermal carbonization (HTC) of biomass to derive hydrochar that has demonstrated key role for soil amendment purposes by increasing the water holding capacity of soil when mixed with it. However, while thinking beyond the scope of Earth, the question that intrigued our research was the possibility of enhancing water retention capability to boost plant growth in outer space to facilitate life there, for example on the International Space Station (ISS) or at human outposts on Mars. In order to limit transport cost while conserving energy, NASA's advanced life support system research is pursuing native materials mined on Mars to be utilized for plant growth in future outpost where terrestrial porous materials might be useful to grow plants onboard during future space missions. Hence, this study stemmed in developing composite from exceptionally porous biomass-derived carbon material and Mars regolith to determine the effect of porous carbon material on the water holding capacity of Mars regolith with different mixture proportions. For this study, loblolly pine was used as biomass precursor where HTC was done at 260°C to derive hydrochar which was then chemically activated using KOH at 800 °C to fabricate exceptionally porous carbon material, named activated hydrochars. Activated hydrochar was then mixed with Mars regolith where mass proportion followed was 0, 5, 10, 25, 50 (wt.%) of activated hydrochar in 1g of Mars Regolith. KOH was then added to the mixture following a mass ratio of 1:4 (mixture: KOH) while adding 25 to 50 ml of water to create a suspension for homogenous mixing during ultrasonication. The mixture was ultrasonicated for 6 hours followed by hot-plate magnetic stirring for 12 hours in order to evaporate the water off prior to activation in tube furnace for a residence time of 2 hours at 800 °C. The activated hydrochar-composites hence produced were characterized by N_2 adsorption–desorption for surface porosity quantification, scanning electron microscopy for morphology appearance analysis, X-ray powder diffraction for crystallinity analysis, proximate and ultimate analysis for detailed chemical composition analysis. Eventually, the applicability of the composites was explored by assessing the water holding capacity where the individual composites were saturated with water while gently agitating until excess water was observed. Then, the water-composite mixture was allowed to rest for 24 hours to facilitate a homogenous mixture with water where the mixture was then allowed to be drained by gravity through filter paper (Whatman 41) for another 24 h. The wet composite mixture was then weighed prior to drying and then again weighed to determine the dry mass from which the water holding capacity of each mixture was determined.

593c Optimization of Pyrolysis Parameters for Biochar Production

Aviv Kresch, David Jiang, Xiao Lin, Amanda Simson

Pyrogenic Carbon Capture and Storage (PyCCS) is a method of carbon sequestration that is gaining popularity as a way to mitigate rising atmospheric CO_2 levels. Biochar production is maximized with slow pyrolysis and low temperatures. Pyrolysis is generally an endothermic process, but some research indicates that there are specific temperatures at which pyrolysis is exothermic. Pyrolysis enthalpy also varies by feedstock. As pyrolysis is conducted on a larger scale, the optimal temperature for pyrolysis to minimize energy use and maximize biochar production becomes increasingly important for both economic and environmental reasons. Pyrolysis time-temperature data at temperatures between 400°C and 800°C for several feedstocks to determine the effect of pyrolysis temperature and feedstock on enthalpy of the reaction. In addition to measuring temperature during pyrolysis, the composition of non-condensable gases was also measured to understand pyrolysis reaction changes as a function of pyrolysis temperature and feedstock.

593d Effect of Surfactant on HTL Derived Hydrochar Physicochemical Characteristics

Khang Huynh, Bharathkiran Maddipudi, Anuradha Shende, Rajesh Shende

Porous carbon (POC) is a versatile material that can be used in various applications such as carbon capture, hydrogen storage, soil amendment, wastewater treatment, energy storage and more. Lignocellulosic biomass (LCB) is a useful, abundant and environmentally friendly feedstock as compared to fossilized feedstocks to produce POC. Thermochemical conversion of LCB via hydrothermal liquefaction (HTL) can be employed to obtain hydrochar. Typically, the hydrochar exhibits poor physicochemical characteristics, which needs activation steps to improve specific surface area (SSA), pore volume and a desirable pore size. Superior physicochemical characteristics are desirable for adsorption and energy storage purposes. Although several chemical agents, such as H_3PO_4, $ZnCl_2$, and KOH known to improve physicochemical characteristics of hydrochar, the effect of surfactants during the chemical activation step is not available. In this study, HTL-derived hydrochar from corn stover was chemically activated with KOH in presence of a surfactant and thermally treated to obtain POC. Three different surfactants, namely Pluronic F127 (non-ionic), cetyltrimethylammonium bromide (CTAB, cationic) and sodium dodecyl sulfate (anionic) were used at different weight ratios during the chemical activation step to study their effects on the characteristics of the derived biochar. It was then thermally treated for 1 hour under constant flow of UHP N_2 gas and acid washed further with 0.1M HCl to obtain POC. These POC samples were characterized by Brunauer-Emmett-Teller (BET) surface area analyzer, Fourier transform infrared (FTIR) spectroscopy, and scanning electron microscopy (SEM). Synthesis and characterization of POC will be presented.

598b Pressurized Steam Gasification of Biomass Char – Influence of Temperature and Gas Composition

Jieun Kim, Kevin Whitty

With increasing interest in renewable energy and reducing greenhouse gases, biomass is one of the fuel sources that can have negative carbon emissions. This study investigates the reaction kinetics of biomass char steam gasification under high pressure and the inhibition effects of syngas, hydrogen and carbon monoxide. The water-gas shift reaction during steam gasification of biomass char produces carbon dioxide and hydrogen. It has been observed that increasing concentration of hydrogen and carbon monoxide inhibit the reactivity of steam gasification. In order to investigate the influence of gas composition on gasification rate, a series of experiments were performed. Pine sawdust was pyrolyzed at 900 °C with a heating rate of 50 °C/min in thermogravimetric analysis (TGA) reactor under different pressures. Then, the biochar went through steam gasification at a temperature ranging from 700 to 950 °C and pressures from 1 to 60 bar. The reactivity at different temperatures, pressure, and compositions of carbon monoxide and hydrogen will be presented. The kinetic model for steam gasification, including these inhibitors, will also be presented.

598c Surfactant Assisted Hydrothermal Liquefaction of Corn Stover

Bharathkiran Maddipudi, Khang Huynh, Paiton Mueller, Rajesh Shende

Worldwide, as corn crop yield is more than 1 bty, significant amount of corn stover feedstock is available globally. In this study, we investigated hydrothermal liquefaction (HTL) of corn stover at 250°C and 800-1000 psi (system autogenous pressure) in presence of ionic and non-ionic surfactant. Presence of a surfactant is known to change the wettability, reduce the hydrophilicity of particles and increase interfacial surface area, which can influence characteristics of hydrochar as well as distribution of liquified products, mainly the C1-C3 carboxylic acids. HTL process resulted in product gas, aqueous biocrude, heavy bio-oil (HBO) and solid char. The biocrude was characterized by TOC analyzer and gas-chromatography mass spectrometry (GC-MS) to understand extent of liquefied carbon and oxygenated hydrocarbons, respectively. Both solid char and HBO were analyzed by the elemental analyzer and HHVs were estimated based on C, H, N and O content. Yield and quality of the products generated from corn stover feedstock will be presented to understand the effect of surfactants during HTL processing.

496a Recent Advances in Intensification of Bio-Based Processes By Ultrasound and Hydrodynamic Cavitation

Yusuf Adewuyi

Over the last two decades, there has been a plethora of studies utilizing ultrasonic and hydrodynamic cavitation to intensify and optimize the homogeneous and catalytic synthesis of biodiesel and the conversion of lignocellulosic biomass to ethanol and other chemicals for biofuels production so as to reduce cost and make the processes economically attractive for large-scale production. Cavitation is the formation, growth and implosive collapse of gas- or vapor-filled microbubbles and can be induced acoustically or hydrodynamically in a body of liquid. Cavitation is known to produce extreme conditions of temperatures and pressures. The cavitation event also gives rise to acoustic microstreaming or formation of miniature eddies that enhance the mass and heat transfer in the liquid. Biodiesel has become one of the premier forms of alternative energy source. Biodiesel is produced by the reaction of lipid feedstocks with an alcohol such as methanol in the presence of homogeneous catalysts as NaOH, KOH or their methoxides, and acids such as H_2SO_4, H_3PO_4; and solid superacid catalysts. The conversion of lignocellulosic biomass for biofuels and biorefinery applications is also limited due to the cost of pretreatment to separate or access the biomass's three main components, cellulose, hemicelluloses, and lignin. Ultrasound and/or cavitation-assisted technologies are green techniques that have the potential to enhance the separation and hydrolysis of lignocellulosic materials for the use in biofuel production and biorefineries, through physical and/or chemical mechanisms. This talk will discuss the role of catalysis and ultrasound-assisted processes in energy sustainability, and economic challenges in the development of green routes for bio-energy and bio-refineries. A synopsis of the some results of our on cavitation-enhanced intensification and optimization of the syntheses of biofuels from edible and non-edible biomass will also be presented.

496b Glycerol As a Co-Solvent for Hydrothermal Liquefaction of Corn Stover

Khang Huynh, Bharathkiran Maddipudi, Anuradha Shende, Rajesh Shende

Hydrothermal liquefaction (HTL) is a thermochemical process capable of converting lignocellulosic biomass (LCB) such as woody biomass, forest/agriculture residue, switchgrass, and corn stover into value-added products. Depending upon the HTL processing conditions used, the distribution of products such as bio-oil, hydrochar, phenolics, C1-C3 carboxylic acids, and oxygenated hydrocarbons would vary. Typically, LCB is converted in a hot, pressurized water for a residence time of 1-2 hours at low to moderate operating temperatures as compared to the pyrolysis. These operating conditions result in low tar yield and high conversion efficiency. The use of glycerol is known to influence product distribution as well as the yield. In this study, HTL of corn stover was carried out at 250-300°C using glycerol as a co-solvent. Biomass to solvent/co-solvent (B:S) weight ratio was varied to study the effect on products yield and their distribution. The resulting liquid product (biocrude) after HTL was fully characterized by elemental analysis and gas-chromatography/mass-spectrometry (GC/MS). The solid product, hydrochar was recovered and characterized by elemental analysis, Brunauer-Emmett-Teller (BET) surface area analyzer and FTIR. The effect of glycerol on products yields and quality was evaluated compared with the products obtained in the absence of glycerol. Experimental results of HTL processing of corn stover in the presence of glycerol will be presented.

496c A Study of Trace Ammonia Impurities in Renewable Natural Gas and Their Impact on Its Potential Utilization

Linghao Zhao, Zhuofan Shi, Jingwen Gong, Jorge Gutierrez, Siari Sosa, Theodore Tsotsis

Renewable natural gas (RNG), produced from biogas upgrading, is an important alternative to fossil fuels for power generation. Unfortunately, RNG contains several trace contaminants. One such contaminant is NH_3, of particular concern for RNG produced from farming operations. The presence of NH_3 in RNG can have serious consequences. These can range from damage to the NG infrastructure, to corrosion of analytical and end-user equipment and added pollutant formation during combustion. We have previously reported in the area of pollutant emissions, where we tested several common end-user equipment. They were all operated with NG injected with NH_3 concentrations ranging from 0 to 500 ppm_v. Emissions studied included unburned hydrocarbons (UHC), CO, and NO_X, the latter being of particular concern. The presence of NH_3 resulted in increased NO_X emissions for all equipment.

In this talk our focus is on NH_3-induced corrosion in materials relevant to the NG infrastructure with the goal to identify various (bulk and surface) species potentially forming, their rates of formation, and their dependence on the operating conditions. The tested materials included components made of steel and brass, and a variety of polymer gaskets. They were exposed to both dry and humidified NH_3/NG mixtures with 0-500 ppm_v of NH_3 for periods, under each concentration, ranging from 400-1200 hours. NH_3-induced changes in the materials were identified via direct weight change measurements and optical observations, SEM-EDS, ICP-OES and microhardness testing. For the metal components, corrosion phenomena were observed, particularly for the higher NH_3 concentrations under humidified conditions. We also report here the impacts of NH_3 on Gas Chromatographs used for customer billing. NH_3 has a detrimental impact on the instrument the severity depending on exposure length. The results of the study are concerning, indicating the need for thorough purification of the RNG prior to its potential injection into the NG network.

496d Multi-Scale Modelling of Low-Temperature Electrolysis of Steam in Molten Carbonate Cells

Maria Anna Murmura, Luca Turchetti, Maria Cristina Annesini

High-temperature electrochemical cells, such as solid oxide and molten carbonate cells, have been recently receiving attention for both electricity generation (fuel cells) and hydrogen or CO production (electrolysis cells). Different scales of models may be found in the literature. 0D models, which allow to describe the correlation between current density and voltage, can be useful to gain some insight on the main phenomena responsible for overpotential in a given cell; however, the results cannot be extrapolated to cells operating under different operating conditions or with different geometries. 1D models allow to describe activation and ohmic overpotentials, but not the effect of concentration gradients of gaseous species participating in the electrohcemical reactions. On the other hand 2D models allow to take into account the different electrochemical and transport phenomena determining the performance of a cell and, once developed, they allow to analyze the effect of changes in operating conditions. Obviously, the difficulty in developing these models goes hand-in-hand with their degree of completeness and versatility. It is interesting to note that 2D models for electrochemical cells are rare in the liteature, although models of similar degrees of complexity exist for chemical reactors in which no electrochemical processes occur. The difficulty in finding thorough models is more pronounced for electrolysis cells than for fuel cells.

Molten carbonate electrolysis cells (MCECs) are currently being considered for the high temperature electrolysis of steam; however, studies on these cells are also very scarce.The MCEC configuration studied consists of two porous electrodes, separated by a molten carbonate electrolytic mixture. Gas is fed to the the cathode and anode sides of the cells and flows through channels running parallel to the electrodes themselves. The model was employed to study the performance of the cell operating both at the high temperatures traditionally proposed for MCECs of approximately 600°C and at temperatures around 500°C. In terms of cell assembly, the main difference when operating at lower temperatures lies in the need to use a different mixture of carbonates, having a melting temperature lower than 400°C and that is therefore liquid under the operating conditions envisaged. Changing the composition of the electrolyte mixture has an effect on the ionic conductivity both in the electrolyte layer and within the porous electrode.

In the present work different models have been developed to analyze the behavior of a planar molten carbonate electrolysis cell. The results were compared with experimental data available in the literature. The 0D model allowed to determine that, under common operating conditions, the performance of these cells is limited by activation and ohmic overpotentials. To gain further insight, a 1D model was developed in which it was possible to distinguish between (i) the anodic and cathodic activation overpotentials and (ii) ohmic overpotentials due to ion transport in the electrolyte present in the cathode, anode, and electrolyte matrix, as well as electron transport in the electrodes. The results revealed that the overpotentials could be mainly attributed to activation at the anode and ohmic losses due to the resistance of ion transport within the electrolyte matrix.

The 2D model developed takes into account

1. Mass transport in the gas channels and in the porous electrodes;
2. Charge transport in the electrodes and electrolytes;
3. Heat transport throughout the cell.

With regards to mass transport, the phenomena considered where convection, diffusion and reaction. In the cathodic gas channel the possible occurrence of reverse-water-gas shift (rWGS) was considered. The electrochemical semi-reactions of hydrogen and oxygen production were accounted for in the cathode and anode, respectively. Charge transport by electrons was described in the solid fraction of the electrodes; whereas charge transport by ions was described in the liquid electrolyte, both within the electrode porosity and in the electrolyte layer itself. As for heat transfer, the non-uniformity of temperature is due to the heat effect of both the rWGS reaction and the electrochemical water-splitting reaction as well as that of Joule heating.

The 2D model explicitly accounts for concentration gradients of the gaseous species in the directions both parallel and orthogonal to the direction of the main gas flow. In so doing, it was found that the activation losses at the anode had been slightly overpredicted by the 1D model.

Overall, it was found that ohmic losses due to ion transport in the electrolyte matrix are high, and should be reduced to improve the performance of the cell. This can be done by acting on the geometric configuration of the cell, so to reduce the distance between the electrodes.

The present work presents, to the best of the authors' knowledge, the first report of multi-scale modelling of MCECs. The results have allowed to determine the factors limiting the behavior of these electrochemical cells and thus provide a guide to the design of more efficicient cells.

592a Biomass Cofiring with Precombustion Carbon Capture Baseline Testing at Und Eerc

Joshua Stanislowski

The University of North Dakota Energy andamp; Environmental Research Center (EERC) performed testing to establish a baseline of precombustion carbon capture performance when combined with coal- and biomass-derived syngas. Goals of the project were to generate high-pressure syngas from a matrix of coal types and biomass types and to vary the concentrations of the fuels. Carbon dioxide was separated at pressure from the syngas stream using a physical solvent. Solvent performance and contaminant accumulation within the acid gas separation system (AGSS) were examined. The work was supported through U.S. Department of Energy National Energy Technology Laboratory funding.

The EERC conducted tests using the high-pressure fluid-bed gasifier. The test also utilized warm-gas cleanup equipment comprising a hot-gas filter vessel, presulfided sour shift catalyst, sulfur sorbents, a quench system, and an AGSS using a physical solvent based on dimethyl ether of polyethylene glycol. The test plan included 15 weeks of testing using three different ranks of coal (North Dakota lignite, Powder River Basin subbituminous, and western bituminous) as baselines and then blended with wood and corn stover at 25% and 50%. Each week of testing was targeted toward a specific coal or coal–biomass blend. Operational data, observations, key findings, and recommendations will be presented.

592b Investigating Oxidative Degradation Products and Mechanisms of Aminopolymer Sorbents for Direct Air Capture (DAC)

Yoseph Guta, Juliana Carneiro, Giada Innocenti, Simon H. Pang, Miles Sakwa-Novak, Christopher W. Jones, Carsten Sievers

Aminopolymer based sorbents have shown promising results in direct air capture (DAC) of CO_2 from ambient air owing to their high CO_2 adsorption capacity and selectivity. While those adsorptive properties are important, the stability of a sorbent is important in creating high-performing, cost-effective and long-lasting sorbents. The high concentration of molecular oxygen (O_2) in ambient air (~21 %) and temperature swings used during sorbent regeneration cause oxidative degradation of these sorbents with time. Either via process upsets that could cause rapid oxidation, or slow oxidation under milder, controlled operating conditions, oxidative degradation has the potential to increase operation costs.

This work investigates the oxidative degradation mechanism(s), pathways, and major products of aminopolymer sorbent reactions with O_2 using a model sorbent (45 wt% poly(ethylenimine) (PEI) on γ-Al_2O_3) under varied experimental conditions. We explore the effect of oxidative degradation on the CO_2 adsorption capacity using thermogravimetric analysis, ATR-IR spectroscopy, High-Resolution Time-of-Flight Chemical Ionization Mass Spectrometer (HR-ToF-CIMS), ^{13}C NMR spectroscopy and elemental analysis (C/N ratio). The results show the formation of low molecular weight amine compounds, while the sorbent deactivation seems to follow a free radical initiated autoxidation, following autoxidation pathway that has been described for a class of polymers such as rubber during thermo-oxidation.

592c Quaternary Ammonium Functionalized Poly(arylene ether sulfone) Copolymers for Direct Air Capture

Hoda Shokrollahzadeh Behbahani, Matthew D. Green

Unprecedented rise in atmospheric CO_2 levels has amplified Earth's greenhouse gas effects and significantly contributed to global warming. Direct air capture (DAC) has the potential to develop into an important tool to keep below the critical temperature rise threshold of 1.5 °C reported in the latest Intergovernmental Panel on Climate Change (IPCC). The moisture swing process is advantageous for DAC because it has low energy requirements. In this study, novel quaternary ammonium functionalized poly(arylene ether sulfone) polymers with bicarbonate (HCO_3^-) counterions with various ion exchange capacities (IEC) are synthesized for direct air capture by the moisture swing process. The backbone of the polymer is allyl modified poly(arylene ether sulfone) copolymer (PAES-co-APAES) synthesized by step growth polymerization, and quaternary ammonium group is added to the backbone through post-polymerization functionalization to make PAES-co-QAPAES[HCO_3^-] random copolymers. Size exclusion chromatography (SEC) is utilized to measure the molecular weight of the novel polymers and the chemical structure and IEC is analyzed through nuclear magnetic resonance (^{1}H NMR) spectroscopy. Thermogravimetric analysis (TGA) showed good thermal stability with 5% degradation temperature ($T_{d5\%}$) above 240 °C. Experimental data show stable atmospheric CO_2 capture and release response with changes in humidity from 8 (30% RH) to 25 ppt (95% RH) at 22 °C for six consecutive cycles.

592d Synthesizing DAC Hydrophobic Polymers with Various Form Factors

Ani Nazari, Matthew D. Green

An increase in the emission of atmospheric CO_2 has caused global concerns. According to a report published by the Intergovernmental Panel on Climate Change (IPCC), it is expected that CO_2 emissions will rise from the current 36 Gt/yr to between 48 and 55 Gt/yr by the year 2050 as energy demand increases between 40% and 150%. Additionally, based on the Paris Climate Agreement in 2015, a limit of 2 °C was set for the global temperature increase while targeting an even lower limit of 1.5 °C. Since emission reduction alone cannot be a solution to climate change, negative emission technologies have gained much attention. Direct Air Capture (DAC) is one of the promising negative emission technologies that can capture CO_2 directly from ambient air and reduce the CO_2 concentration in the atmosphere. Thus, to achieve that, sorbents with high capturing capacity, low cost and fast kinetics are required. The current study investigates a series of amine-based DAC polymers with hydrophobic characteristics utilized in the humidity-driven CO_2 capture setup. We synthesized series of [2(Methacryloyloxy)ethyl]trimethylammonium chloride (TMAEMA-Cl) , Di(ethylene glycol) dimethacrylate (DEGDMA) and 2,2,2-Trifluoroethyl methacrylate (TFMA) membranes with various ion exchange capacities. These sorbents with quaternary ammonium cations on their backbone and bicarbonate counterions absorb CO_2 when exposed to ambient air and release it when in contact with an aqueous solution. The capturing process is driven by the energy generated from the evaporation of water. To further investigate the capturing capacity of the hydrophobic sorbents, they were ground into powder. The powdered sorbents were then added to the surface of an adhesive and tested in moisture-swing set up for different cycle times. The capturing capacity and kinetics of the two form factors were then investigated for various ion exchange capacities. Herein, we studied designing and synthesizing flexible and porous sorbents with high stability and fast kinetics to be used in the moisture-swing CO_2 capture mechanism and the effect of form factor on the kinetics of the sorbents.

592e Sorbent Regeneration Energy Analysis of Phase-Changing Guanidine-Based Ligands Used for CO_2 Direct-Air Capture

Abishek Kasturi, Sotira Yiacoumi, Diana Stamberga, Radu Custelcean, Costas Tsouris

Currently, solvents that are used for CO_2 capture face issues regarding large energy penalties associated with regeneration and water vaporization. Bis-imino-guanidine (BIG)-based ligands, when combined with potassium salts of amino acids (e.g., potassium sarcosinate, K-SAR), react with CO_2 to form an insoluble precipitate. Research conducted with BIG ligands suggests that this phase change has the potential to reduce the regeneration energy requirement to levels below that of current industry standards for direct air capture (calcium carbonate, $CaCO_3$) and flue gas capture (monoethanolamine, MEA). This study investigates the thermal regeneration of sorbents used for the direct air capture of CO_2 using spectroscopic and calorimetric methods. In particular, the regeneration energy requirement for 3M K-SAR and a guanidine-based ligand (MGBIG) is quantified using differential scanning calorimetry (DSC) coupled with thermogravimetric analysis (TGA) and Fourier-transform infrared (FTIR) spectroscopy. In order to fully understand the different thermodynamic parameters contributing to the regeneration of K-SAR and MGBIG, the sensible heat, enthalpy of reaction, and enthalpy of vaporization are also estimated. DSC/TGA helps with understanding the regeneration from a thermodynamic perspective and provides information regarding the heat flow as a function of mass, time, and temperature. The gas generated during thermal regeneration is analyzed using FTIR, which provides real-time compositional information of the evolved gas as a function of reaction time and temperature. Using these three techniques, the total regeneration energy of 3M K-SAR and MGBIG was approximated to be 21.4 GJ/tCO_2 and 7 GJ/tCO_2, respectively. The enthalpy of reaction accounts for approximately 11% of the total regeneration energy for K-SAR and approximately 77% of the total regeneration energy for MGBIG. These results provide the fundamental basis for developing an effective carbon capture technology with phase-changing amino acid/guanidine absorbents that can be used to effectively capture atmospheric CO_2.

This research was funded by the US Department of Energy, Office of Technology Transitions, through a Technology Commercialization Fund (TCF-20-20118) supported by the Office of Fossil Energy and Carbon Management.

592f Effect of Fundamental Biopolymers in the Development of Surface Porosity and CO_2 capture Capacity of Superactivated Hydrochars

Al Ibtida Sultana, Toufiq Reza

Recently, hydrothermal carbonization (HTC) of biomass has been extensively studied to derive surface functionality rich hydrochar that can be chemically activated (CAC) to create porous superactivated hydrochars (SAH) for gas storage applications like carbon dioxide capture. However, despite identical thermochemical conversion (HTC and CAC) conditions, the SAHs derived from different biomass precursors exhibit dissimilarity in surface morphology as well as its corresponding gas storage capacity. Henceforth, finding the fundamental key underlying this discrepancy was of particular interest and motivated this study. For example, food-waste and loblolly-pine-derived SAHs carbonized at 220°C and activated at 800°C demonstrated BET surface area of 2885 and 1531 m^2/g respectively. The distinct disparity in surface area puts special emphasis in rather revealing the potency of individual biopolymer constituents of a biomass in the process of favorable pore evolution that eventually affects CO_2 captured. Therefore, this study was designed to explore the biopolymer model compounds of cellulose, hemicellulose, lignin which were independently hydrothermally carbonized at 260°C and then activated using KOH to synthesize SAHs at temperatures using KOH: hydrochar ratio of 2:1 and 4:1. Furthermore, the synergistic effect of the biopolymers were also evaluated by proportionately formulating into simulated loblolly pine which also underwent HTC and CAC at aforementioned process conditions. All SAHs were characterized by N_2 adsorption–desorption, scanning electron microscopy, X-ray powder diffraction, proximate and ultimate analysis. Eventually, the applicability of the SAHs were explored by capturing CO_2 at ambient condition using High Pressure Volumetric Analyzer. The conclusion from our study could serve as the preliminary basis of establishing role of various fundamental biopolymers in pore evolution and hence preparation of high-performance biomass-based CO_2 storage medium.

592g Performance Evaluation of Membrane Module for CO_2 Separation from Ambient Air

Palash Panja, Maksudul Alam, Milind Deo

Membrane technology has been used since the early sixties for gas separation. Some industrial applications include the separation of methane from biogas, separation of nitrogen or oxygen from the air, and recovery of hydrogen from ammonia plants and other chemical processes. A membrane is synthesized for specific gas separation; hence types and characteristics of the membrane vary depending on its applications. Membrane technology could be a promising method to separate CO_2 from ambient air because a traditional method like the amine-based solvent absorption method is not suitable for ultra-low concentrations of CO_2 (i.e., ~400 ppm). Non-porous polymeric membranes where molecular diffusion is the dominant mechanism of mass transfer are more effective for this purpose.

The entire mechanism of the gas transport from upstream to downstream of polymeric membrane consists of three steps (i) adsorption on the upstream boundary (ii) diffusion through membrane matrix and (iii) evaporation from the downstream boundary. The mechanism of transport of gas molecules through the membrane depends on the structure and material of the membrane. The overall module performance is dependent on operating conditions (such as pressure difference), selectivity of gases, and membrane area and thickness.

In this study, carbon dioxide (400 ppm) is separated from a binary mixture with nitrogen to reduce the amount to 300 ppm which is considered as the concentration of CO_2 in the air during the pre-industrial era. As a result, the concentration of CO_2 in the permeate stream is enhanced to many folds which can be utilized for other purposes. Separation performance is evaluated based on the pressure ratio between the upstream and downstream, membrane parameters (area and thickness), and the permeability of components or selectivity. The membrane thickness of 1 μm is used. The pressure ratio of 55 (upstream pressure: 110 kPa, downstream pressure: 2 kPa) is kept constant for all simulations. The single-stage separation process with the co-current flow and constant permeability flux model is simulated to assess the sensitivity of selectivity which is the ratio of the permeance of two components.

The flow rates and concentrations of all components in the retentate and permeate are obtained from the simulations for different conditions. The separation is influenced by the membrane area when other parameters are fixed. It is often practical to use stage cut which is defined as the ratio of total permeate flow rate and feed flow rate instead of membrane area to compare different simulation results on the same scale. It is noted that the stage cut varies between 0 to 1 with the membrane area. The purity of CO_2 is also analyzed which is defined as the concentration of CO_2 in permeate. Another parameter is the amount of CO_2 in permeate which is quantified by the recovery factor (ratio of CO_2 in permeate and CO_2 in feed).

In this co-current flow scheme, the retentate stream is in equilibrium with permeate stream on the other side of the membrane. The CO_2 concentration in the retentate and permeate decreases along the membrane. However, the total flow rate also increases thus increasing the stage cut as a result and enhancing the CO_2 recovery factor. This is a trade-off situation between the purity and recovery factor of CO_2 in the permeate stream. This can be fixed by setting the CO_2 concentration in the retentate to a certain value such as 300 ppm. Therefore, the membrane area is designed to achieve a specific amount of separation. For a fixed amount of separation, the higher CO_2 concentration in the permeate is achieved for higher selectivity. For example, CO_2 concentration in the permeate is 685 ppm for selectivity of 2 and 13601 ppm for selectivity of 200. The CO_2 recovery factor doesn't change much for selectivity above 20. It is 44% for selectivity of 2 and then it reaches a plateau of 26% for selectivity above 15. For fixed CO_2 concentration in the retentate (i.e., 300 ppm), the stage cut is lower and the CO_2 concentration in the permeate is higher for the higher selectivity. As a consequence, the recovery factor which is the combination of these two for fixed feed rate and compositions remains constant. Therefore, the high-selectivity non-porous polymeric membrane with a low surface area is an effective technology to separate CO_2 from the air to combat global warming

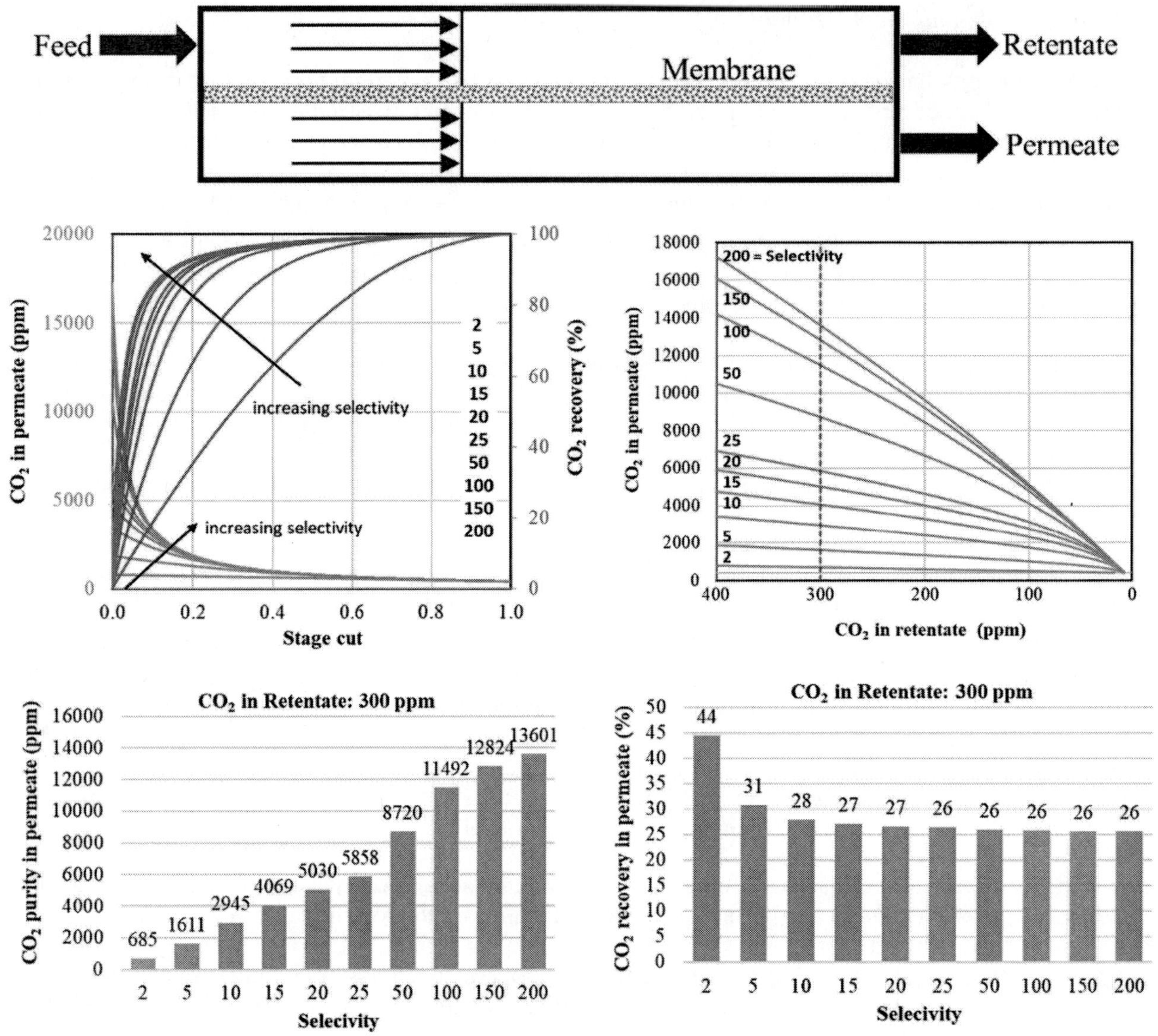

628a Techno-Environmental Analysis of Different Pre-Combustion Carbon Capture Technologies for Hydrogen Production from Waste

Alex Sebastiani

1. Introduction

Recently, hydrogen from low-carbon routes, has garnered attention as a high-density energy vector with low greenhouse-gas production emissions and no emissions at its point of use. The proposed sustainable route to produce Hydrogen is the gasification of biomass waste feedstock coupled with pre-combustion capture and long-term geological storage of carbon dioxide. This system, known as Bioenergy with carbon capture (BECCS), effectively removes CO_2 from the natural carbon cycle and is considered a technology that will play a pivotal role in keeping global warming well below the recommended 2°C since the pre-industrial period.

This research work focuses on the techno-environmental analysis of a complex product system that a) produces hydrogen as the main product b) captures and permanently sequesters carbon dioxide, a by-product and c) utilises municipal solid waste (MSW) feedstock, thereby diverting its fate from landfill or incineration.

This research highlights the use of biomass waste, resulting in lower costs and added environmental benefits, and its efficacy of producing grid quality hydrogen. However, the central theme of this work is an investigation into carbon capture technologies for the Waste-to-H_2 process. These include four that are already commercially available – i.e., amine scrubbing via monoethanolamine (MEA), a Benfield-like process, which uses potassium carbonate as solvent, Selexol® and Rectisol® - and Sorption-Enhanced Water Gas Shift (SEWGS). MEA and Potassium carbonate rely on chemical absorption. Selexol® and Rectisol® use physical solvents (dimethyl ethers of polyethylene glycol (DEPG) and refrigerated methanol respectively). SEWGS is a novel low-TRL technology that combines the water–gas shift (WGS) reaction with the selective adsorption of CO_2 on a potassium promoted hydrotalcite sorbent at high temperature and pressure.

2. Materials and methods

A sophisticated set of models to predict the performance of different pre-combustion CCS technologies, as well as the balance of plant on a fully integrated biohydrogen (Bio-H_2) system was developed using Aspen Plus simulation software. The facility is designed and modelled to convert approximately 110,000 tonnes per annum of municipal solid to approximately 50 MWh of grid-quality hydrogen. Upon collection, the waste is sorted and pre-treated. The obtained refuse derived fuel (RDF) is fed into a two- stage process for steam-oxygen gasification and tar-reforming to produce a clean syngas. The hydrogen fraction of the clean syngas is then increased via a series of catalytic water gas shift reactors. The carbon dioxide is separated using typical carbon-capture technologies and is then liquefied for permanent storage. Pressure swing adsorption (PSA) is employed to achieve the purity required for H_2 use for domestic heating, while the tail gas is burnt in a gas engine to generate heat and electricity.

Following the guidelines of the ISO 14040 and ISO 14044 standards, a Life Cycle Assessment (LCA) methodology was applied to investigate the environmental and carbon performance of greenhouse gas removal via capture of biogenic CO_2 from waste, while producing Bio-H_2. The multi-functionality of such a system lends itself to complexities arising from the choice of system boundaries, functional unit, and assumptions to integrate mature scale elements of the process with other sections at a lower technology readiness level. The function unit chosen here was 1 tonne of grid-quality Bio-H_2 produced.

3. Preliminary Results and discussion

An extensive analysis of the system was performed to assess the most appropriate carbon capture andamp; storage technology for a BECCS system that produces hydrogen from waste. To this end two system boundaries (one focusing on stages affected by the technology change and another considering the complete plant) and different functional units were analysed. Although assumptions and results varied based on functional unit chosen, the ranking of technologies remained the same. Direct emissions are released on-site. Indirect emissions are associated with the supply chain and end-of-life treatment (where applicable) of chemicals and energy in UK. The negative climate change impacts represent avoided greenhouse gas emissions associated with the generation of excess electricity and the permanent sequestration of CO_2 from biogenic sources. This is related to the biogenic fraction of waste that varies between 50 and 80% in weight. The results shown are given in terms of 1 tonne of grid-quality Bio-H_2 from the stages of wet MSW fed into the plant to the generation of Bio-H_2 and compression of CO_2, considering an average 65% biogenic carbon.

Carbon capture technologies based on physical adsorption (i.e. Selexol, Rectisol and SEWGS) outperform those technologies based on chemisorption (MEA and Potassium carbonate) in the climate change category. For chemisorption, high thermal energy requirements in the carbon capture stage to regenerate the solvent is the major contributor to climate change. Similarly, gas compression is the largest contributors for physical adsorption technologies. Major emission savings are realised due to the permanent sequestration of direct CO_2 process emissions.

The results obtained for the climate change category are corroborated by a comparative analysis across all environmental categories. Overall, MEA and Potassium carbonate solvents underperform while SEWGS consistently generates the lowest environmental impacts.

4. Conclusions

The work highlights the importance of BECCS as a negative emissions technology and the further environmental benefits of using a physical absorption technology for carbon capture, particularly SEWGS. This research work provides a real-world application of LCA to a BECCS system producing Bio-H_2 from waste and addresses some methodological complexities of prospective LCAs in order to progress from pilot/demo scale towards commercial scale production.

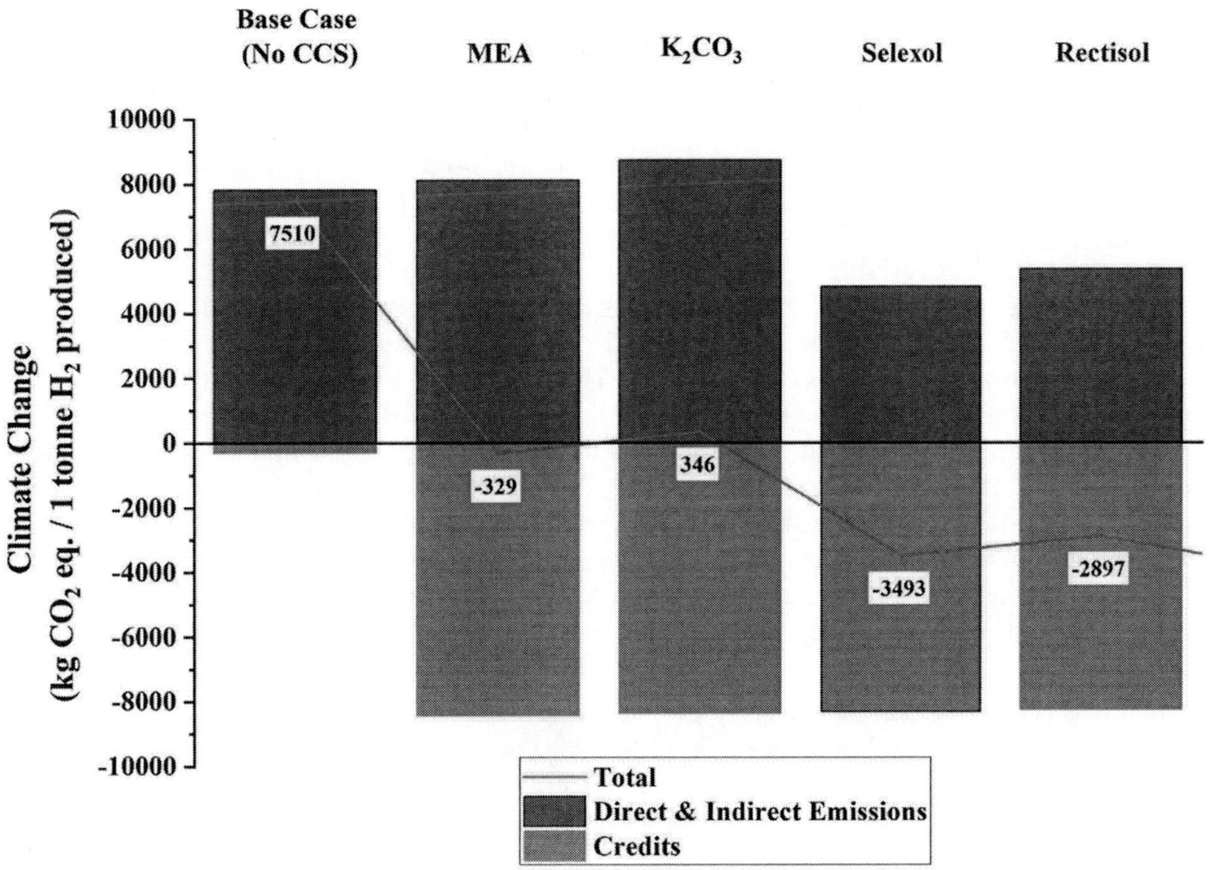
Base Case (No CCS)
MEA
K2CO3
Selexol
Rectisol
Climate Change
(kg CO2 eq. / 1 tonne H2 produced)
10000
8000
6000
4000
2000
0
-2000
-4000
-6000
-8000
-10000
7510
-329
346
-3493
-2897
Total
Direct & Indirect Emissions
Credits

628b Incorporation of Market Signals for the Optimal Design and Operation of a Flexible Post-Combustion Capture System

Radhakrishna Tumbalam Gooty, Jaffer Ghouse, Quang Minh Le, Bhurisa Thitakamol, Sabereh Rezaei, Denis Obiang, Raghubir Gupta, S. James Zhou, Debangsu Bhattacharyya, David Miller

Fossil fuel-based power generators will be required to be more responsive to grid conditions with increasing penetration of variable renewable energy (VRE) generators. Post-combustion carbon capture (PCC) technologies with fast response will need to be deployed to mitigate the carbon emissions from such fossil-based generators. In this work, we evaluate the net present value (NPV) of retrofitting an existing natural gas combined cycle (NGCC) unit with a flexible PCC system while incorporating market signals from a high VRE grid. We use the Los Angeles Department of Water and Power's (LADWP) NGCC unit as a representative of existing NGCC units, and the Svante rapid-temperature swing adsorption (TSA) system for PCC unit. Because of its ability to startup/shutdown and ramp-up/ramp-down in a short time, the chosen capture technology is very attractive for load-following operations. For a given market signal, we formulate a multi-period optimization problem, under a price-taker assumption, to simultaneously optimize the design of the capture system and the operation of the entire plant. Rigorous models for the LADWP NGCC configuration, Svante's TSA system, and the CO_2 compression train are implemented in Aspen Plus Dynamics, Aspen HYSYS, and Aspen Custom Modeler, respectively, and validated with operating data. However, for computational tractability of this large-scale optimization problem, we use surrogate/reduced-order models in the NPV optimization problem. The surrogate model for the NGCC plant is constructed by linearization under nominal conditions, while data-driven nonlinear surrogate models for the capture and compression systems are constructed using the simulation data obtained from the rigorous models. The multiperiod optimization problem is formulated as a mixed integer bilinear programming problem (MIBLP) using IDAES [1] and solved to global optimality using Gurobi 9.5. The solution of the MIBLP yields the optimal size of the capture system and the optimal operating schedule maximizing the net present value. Using this MIBLP formulation, we investigate the profitability of retrofitting an existing NGCC unit located in five different regions, viz. CAISO, ERCOT, MISO-W, NYISO, and PJM-W, under multiple carbon tax scenarios. The results show that the optimal decision strongly depends on the region and on the carbon tax, thereby demonstrating the importance of the inclusion of market signals in the design process. Finally, we determine CAPEX and OPEX reduction targets which would make the inclusion of the capture system more profitable in most of the scenarios. Although we focus on the TSA-based capture system, the optimization framework developed in this work can be used to evaluate any carbon capture technology by suitably modifying the representative models for the capture system.

Disclaimer:

This report was prepared as an account of work sponsored by an agency of the United States Government. Neither the United States Government nor any agency thereof, nor any of their employees, makes any warranty, express or implied, or assumes any legal liability or responsibility for the accuracy, completeness, or usefulness of any information, apparatus, product, or process disclosed, or represents that its use would not infringe privately owned rights. Reference herein to any specific commercial product, process, or service by trade name, trademark, manufacturer, or otherwise does not necessarily constitute or imply its endorsement, recommendation, or favoring by the United States Government or any agency thereof. The views and opinions of authors expressed herein do not necessarily state or reflect those of the United States Government or any agency thereof.

References

1. Lee, A., Ghouse, J.H., Eslick, J.C., Laird, C.D., Siirola, J.D., Zamarripa, M.A., Gunter, D., Shinn, J.H., Dowling, A.W., Bhattacharyya, D., Biegler, L.T., Burgard, A.T., and Miller D. 2021. The IDAES process modeling framework and model library—Flexibility for process simulation and optimization. Journal of Advanced Manufacturing and Processing, 3(3), p.e10095.

628c Optimisation of Carbon Capture and Storage Infrastructure in Wyoming to Reduce Carbon Emissions and Maximise Profits

Matthias Mersch, Niall Mac Dowell

The US state of Wyoming provides a unique environment for carbon capture and storage (CCS). It combines many large carbon emitter sites, including coal-fired power plants, natural gas processing plants and chemical industry, with a great geological storage potential and enhanced oil recovery (EOR) opportunities. Additionally, public and policy support are significant, including monetary incentives in form of 45Q tax credits. In this work, we apply a numerical optimisation model to identify optimal CCS infrastructure, including capture and storage plants as well as CO_2 pipelines, to reduce carbon emissions and maximise profits from 45Q tax credits and EOR.

Wyoming's economy heavily depends on the usage and export of fossil fuels and generated energy. Numbers from the US Bureau of Economic Analysis (BEA) show that in 2020 mining, quarrying and oil and gas extraction is the single largest contributor to the state's gross domestic product (GDP) at 7,923 $M\$_{2012}$, representing 22 % of the total GDP [1]. According to the US Bureau of Labor Statistics (BLS), in February 2022 15,600 jobs were directly associated with the mining and logging industry in Wyoming [2]. CCS is an opportunity to reduce the significant emissions associated with these sectors, making it viable for them to continue operating in a low-carbon future. For this reason, local public and policy support for CCS is high, and significant research has already been undertaken to determine feasibility of capture and storage sites [3] [4] [5]. Additionally, the industry in Wyoming already has experience with carbon infrastructure and EOR. A CCS unit has been operating at Exxon Mobil's LaBarge gas processing facility since 1986 to separate CO_2 from natural gas. The CO_2 is then directed into pipelines and sold to EOR customers [6]. It was recently announced that the carbon capture capacity at that site will be extended [7].

Due to revenues from 45Q and EOR, CCS has the potential to be economically viable. However, the capture, transport and storage infrastructure need to be planned carefully in order to reduce costs and maximise profits. We present a CCS value chain optimisation model that can identify optimal investments into CCS infrastructure.

The proposed model is a spatially and temporally resolved capacity-expansion model. Starting from today's infrastructure, investment into capture plants, storage sites and pipeline networks are optimised on an annual basis until 2050. The target thereby is to maximise profit, defined as difference between revenues from 45Q and EOR and costs, including capture plant, storage well and pipeline investment costs as well as operational expenses. Key constraints include CO_2 availability at emitter sites, maximum storage capacities and injectivities as well as maximum annual pipeline build rates. The model is implemented as mixed-integer linear programming problem in Python/Pyomo and solved by Gurobi.

The model can operate with a variable spatial resolution, as default a grid with 10x10 km resolution is used. Inputs are locations of emitter and storage sites; the optimal pipeline network is a model output. It is important to note that the model does not simply connect all emitter and storage sites but identifies the most economical network and operation. We consider all emitter sites from EPA FLIGHT data [8] with annual CO_2 emissions greater than 0.1 $MtCO_2$, 42 in total. In terms of storage sites, we consider the two geological storage projects that have been explored, CarbonSAFE and Rock Springs Uplift, as well as 15 potential EOR sites with capacities greater than 20 $MtCO_2$. We also consider pre-approved pipeline corridors from the Wyoming Pipeline Corridor Initiative [9], where installation costs are reduced.

First results show that with current 45Q support levels, limited CCs deployment is profitable, focussing on emitter sites with low carbon capture costs such as gas processing plants, and EOR storage sites. Higher 45Q support, as proposed by lawmakers, results in a widespread rollout of CCS across the state and significant CO_2 emission reductions.

The results from this work are not only relevant to identify the most economical CCS infrastructure for given support levels but can also help identify required support levels for defined CCS rollout and/or emission reduction targets. We therefore believe that this research is relevant for a broad range of stakeholders, including industrial players as well as local and central government policymakers.

References

[1] Bureau of Economic Analysis (BEA), "Regional Data - GDP and Personal Income," 31 March 2022. [Online]. Available: https://apps.bea.gov/iTable/iTable.cfm?reqid=70andamp;step=30andamp;isuri=1andamp;major_area=0andamp;area=56000andamp;year=2020andamp;tableid=512andamp;category=1512andamp;area_type=0andamp;year_end=-1andamp;classification=naicsandamp;state=0andamp;statistic=1,2,3,4,5,6,7,8,9,10,11,12,13,14,15,16,17,18,19,20,21,22,23,24,25,26. [Accessed 04 April 2022].

[2] Bureau of Labor Statistics (BLS), "Economy at a Glance - Wyoming," 01 April 2022. [Online]. Available: https://www.bls.gov/eag/eag.wy.htm. [Accessed 04 April 2022].

[3] Enhanced Oil Recovery Institute, "Wyoming is CCUS Ready!," 02 December 2021. [Online]. Available: https://storymaps.arcgis.com/stories/635fa72880ae4d898157b27bd1995ad0. [Accessed 06 April 2022].

[4] University of Wyoming, "Wyoming CarbonSAFE Project," [Online]. Available: https://www.uwyo.edu/cegr/research-projects/wyoming-carbonsafe.html. [Accessed 06 April 2022].

[5] F. McLaughlin and K. Coddington, "Integrated Pre-Feasibility Study of a Commercial-Scale CCS Project in Formations of the Rock Springs Uplift, Wyoming," 2017.

[6] MIT - Carbon Capture andamp; Sequestration Technologies, "LaBarge Fact Sheet: Carbon Dioxide Capture and Storage Project," [Online]. Available: https://sequestration.mit.edu/tools/projects/la_barge.html. [Accessed 06 April 2022].

[7] Reuters, "Exxon plans to increase carbon capture at Wyoming facility," 21 October 2021. [Online]. Available: https://www.reuters.com/business/energy/exxon-plans-increase-carbon-capture-wyoming-facility-2021-10-21/. [Accessed 06 April 2022].

[8] Environmental Protection Agency (EPA), "Facility Level Information on GreenHouse gases Tool (FLIGHT)," 07 August 2021. [Online]. Available: https://ghgdata.epa.gov/ghgp/main.do#/listFacility/. [Accessed 06 April 2022].

[9] Bureau of Land Management (BLM), "BLM National NEPA Register," 20 August 2021. [Online]. Available: https://eplanning.blm.gov/eplanning-ui/project/1502028/590. [Accessed 06 April 2022].

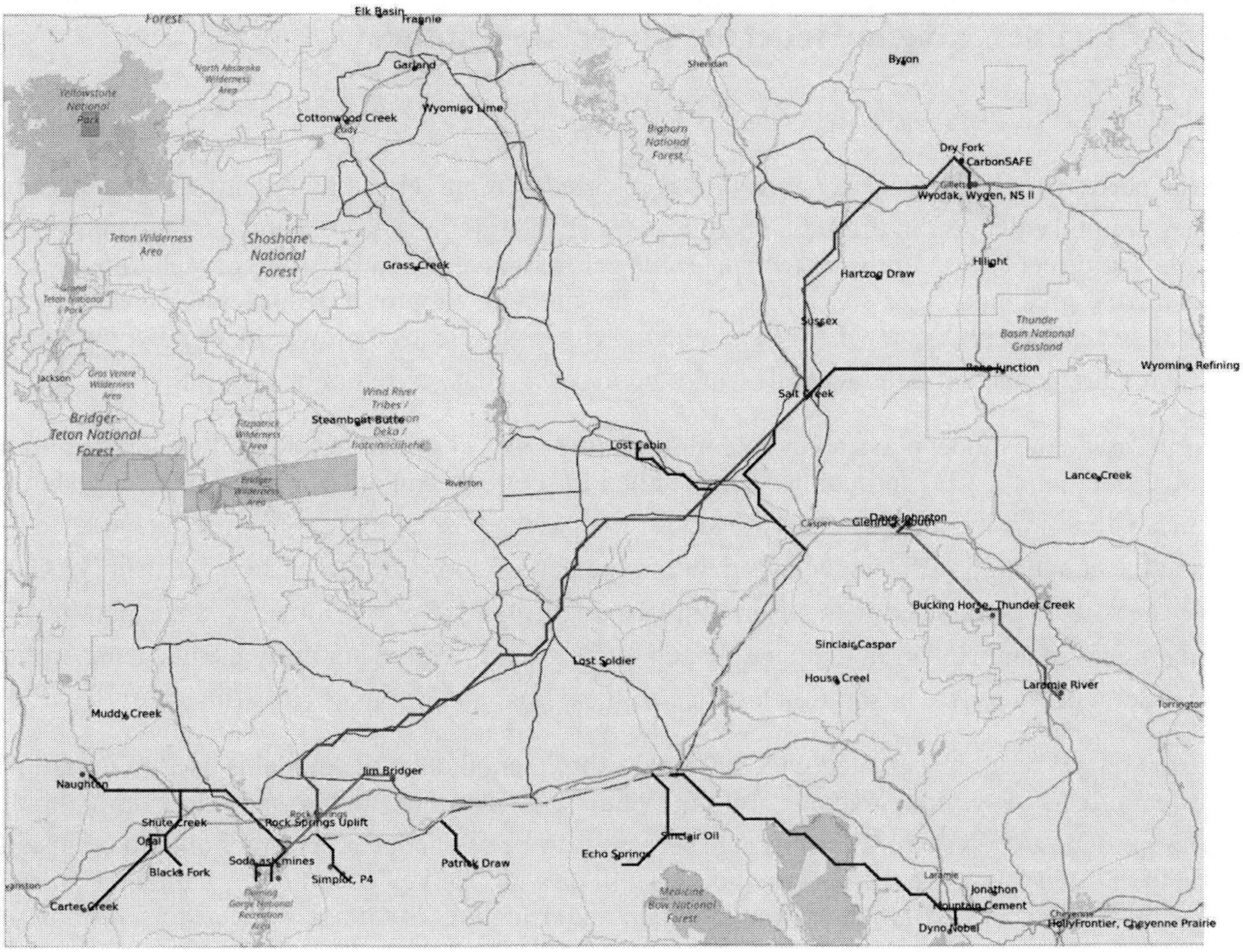

Elk Basin
Frannie
Garland
Cottonwood Creek
Cody
Wyoming Lime
Yellowstone National Park
North Absaroka Wilderness Area
Teton Wilderness Area
Shoshone National Forest
Grass Creek
Sheridan
Byron
Bighorn National Forest
Dry Fork
CarbonSAFE
Gillette
Wyodak, Wygen, NS II
Hartzog Draw
Hilight
Sussex
Thunder Basin National Grassland
Wyoming Refining
Salt Creek
Jackson
Gros Ventre Wilderness Area
Bridger-Teton National Forest
Steamboat Butte
Fitzpatrick Wilderness Area
Bridger Wilderness Area
Riverton
Lost Cabin
Lance Creek
Casper
Dave Johnston
Bucking Horse, Thunder Creek
Sinclair Caspar
House Creel
Lost Soldier
Laramie River
Torrington
Muddy Creek
Naughton
Jim Bridger
Shute Creek
Rock Springs
Rock Springs Uplift
Opal
Blacks Fork
Soda ash mines
Simplot, P4
Patrick Draw
Echo Springs
Sinclair Oil
Carter Creek
Flaming Gorge National Recreation Area
Medicine Bow National Forest
Laramie
Jonathon
Mountain Cement
Dyno Nobel
Cheyenne
HollyFrontier, Cheyenne Prairie

628d Adsorption Kinetic Model for Direct Air Capture

James Akinjide, Aashish Priye, Joo-Youp Lee

A large portion of CO_2 in ~5.2 gigatons (Gt) is released in relatively small quantities from distributed sources emitted each year in the U.S. Therefore, for such emissions, point source CO2 capture is not feasible and direct air capture is an indispensable part of a diversified portfolio of technologies to mitigate U.S. greenhouse gas emissions. Successful direct air capture (DAC) technologies are required to separate high purity CO_2 (e.g., >95%) with high selectivity toward CO_2, low regeneration energy requirements, minimum chemical and thermal degradation, reliability, long lifetime, etc.

Adsorption kinetic model plays a key role in designing DAC systems. In this study, an adsorption kinetic model for our CO_2 sorbent has been studied for DAC potential by integrating CO_2 adsorption kinetics with a flow model. The mass transfer of CO_2 in the porous sorbent bed can be modeled by solving the convection-diffusion equation within the porous sorbent. Intraparticle diffusion of CO_2 inside the sorbent is modeled with adsorption kinetics. CO_2 adsorption isotherm data is obtained for 400 ppm CO_2 in ambient air at temperatures up to 40 °C and relative humidity using chemisorption. The model will couple the adsorption kinetics with external and intraparticle diffusion.

After the model result is validated with lab-scale experimental data, the model is used to investigate the direct effects of system parameters on the overall CO_2 capture efficiency and flow throughput for the design of a DAC system. Currently, the sorbent is to be washcoated onto monolith to be installed inside the DAC system. This model predictions will provide the key design information on overall CO_2 capture efficiency, volumetric productivity, and throughput in terms of dimensions of monoliths including pitch sizes.

628e Design of Passive Contactor Systems for Direct Air Capture

Xin Zhang, WonHee Lee, Sayan Banerjee, Christopher W. Jones, Ryan P. Lively, Matthew Realff

Direct air capture (DAC) is a negative emission technology (NET) for removing CO_2 from the atmosphere to maintain the CO2 level within a reasonable range so as to address greenhouse effects. 3D printing technology has been widely used in producing devices in consequence of the flexible design and rapid prototyping. In this project, 3D printed (3DP) contactors loading with polyethylenimine (PEI) are employed for passive, wind-driven, DAC. Multidimensional computational fluid dynamics (CFD) models are developed to explore the effects of feed wind speed, contactor geometry and porosity distribution, and contactor layout on the wind bypass around the contactors. A 2D process model is then developed by coupling momentum transfer, mass transfer and heat transfer to conduct process design and optimization. Optimization of adsorbent capacity and enthalpy, contactor design, and operating conditions of adsorption and desorption cycles is important to minimize the overall capital and operating costs of the passive contacting systems while reaching a promising productivity. The results of this optimization will be discussed in the talk.

628f Process Analysis for Co-Removal of Methane and Carbon-Dioxide from Air

Devesh Sathya Sri Sairam Sirigina, Aditya Goel, Shareq Mohd Nazir

Background and motivation

The Sixth Assessment Report of the IPCC Report [1] has clearly highlighted the contribution and the effects of non-CO_2 greenhouse gases on global warming. CH_4 is the second largest contributor to global warming after CO_2, with a concentration in the atmosphere that has increased by 63 ppb from 2011. In any scenario where the global average temperature rise is limited to 1-1.9 °C by the end of 2100, these emissions must be significantly reduced (>50%). However, mitigating CH_4 emissions in challenging since majority of the emissions are from distributed artificial and natural sources, and the concentration of it is very low (<2 ppmv) in the atmosphere. Agricultural sector is the major emitter of methane accounting for ~40% of the emissions in the world [2]. CH_4 emissions from the agricultural are more concentrated that the concentration in atmosphere. For example, the concentration of CH_4 in the air from ventilation stables is 10-300 ppmv, and the is 140-25000 ppmv in the headspace of enclosed manure storage [3]. These concentrations are very similar to the concentration of CO_2 in atmosphere. Therefore, synergies can be mapped between direct air capture (DAC) of CO_2 and methane removal from air. This study presents analysis of processes to remove CH_4 and CO_2 from air.

Two routes have been proposed for CH4 removal: (i) separation, capture and regeneration (ii) catalytic (thermal- or photo-) conversion. This work focuses on the second route, where CH_4 is first catalytically converted to CO_2 and H_2O. In the next step, the CO2 formed from catalytic conversion of CH_4 and the existing CO_2 in air is captured in a DAC unit. A first of its kind integrated process to co-remove CH_4 and CO_2 was presented by Sirigina, Goel [4]. In this process, air from ventilation stable having 300 ppmv of CH_4 is blown into a de-humidifier. The de-humidified air is pre-heated to 330 °C and is sent to the catalytic reactor where CH_4 is converted to CO_2 and H_2O. Heat from the air leaving the reactor is recovered before the air enters the DAC unit for CO_2 capture. For an air stream containing 300 ppmv CH_4 at the inlet of de-humidifier, the total energy to remove 1-ton CO_2-equivalent is 9.09 GJ. The energy consumption can be significantly reduced is photo-catalytic reactor designed at room temperature is used to convert CH_4. In this work, we present the comparison of thermal- and phot-catalytic routes to convert CH_4 followed by removing CO_2 in a DAC system.

Methods and assumptions

Air containing 300 ppmv is assumed in the analysis. 6.5 wt% Pd/Al_2O_3 has been considered as the thermal catalyst in this study and the kinetics has been obtained from Alyani and Smith [5]. The de-humidifier is important to the process since the amount of thermal catalyst is sensitive to the water vapor concentration in the feeds stream. The reactor is assumed to be isothermal. The 0D reactor model is developed in Python. 100% conversion of CH_4 is assumed in the reactor. 2% pressure is assumed in the heat-exchangers and de-humidifier. 5% pressure drop is assumed in the reactor and DAC system respectively. A vacuum temperature swing adsorption (VTSA) process is assumed for DAC, with energy requirements obtained from [6]. In the photocatalytic route, the reaction occurs at room temperature (25 °C). The pressure drops in the process are similar to the thermal-catalytic route.

The key performance indicator in this study is the energy (GJ) per ton CO_2-equivalent removed from the process. Global warming potential of CH_4 is assumed to be 25.

Results

The thermal-catalytic route for CH_4 conversion followed by DAC for CO_2 utilizes 9.09 GJ per ton of CO_2-equivalent removed from air. This is very similar to the energy consumed in the DAC processes designed for only CO_2 removal [6]. However, removal of CH_4 has added climate benefits, since CH_4 is a more potent greenhouse gas than CO_2. The energy consumed for co-removal of CH_4 and CO_2 can be significantly lowered through the photo-catalytic route that does not require pre-heating of the feed stream to the reactor. In this study, we compare the performance in terms of energy per ton CO_2-equivalent removed of process routes integrated with thermal- and photo-catalytic reactors. In addition, sensitivity studies for different concentration of CH_4 in feed stream and pressure drop is presented. Results will be summarized in the article.

Acknowledgements

The work is part of the project "Energieffektiv negativa utsläpp frånjordbrukssektorn" (project number 50340-1) (https://www.mgm-negaf.proj.kth.se/) funded by Energimyndigheten (Swedish Energy Agency). This work is also financially supported by the Swedish governmental initiative StandUp for Energy.

References

1. IPCC, Summary for Policymakers. In: Climate Change 2021: The Physical Science Basis.

Contribution of Working Group I to the Sixth Assessment Report of the Intergovernmental Panel on Climate

Change [Masson-Delmotte, V., P. Zhai, A. Pirani, S. L. Connors, C. Péan, S. Berger, N. Caud, Y. Chen, L.

Goldfarb, M. I. Gomis, M. Huang, K. Leitzell, E. Lonnoy, J.B.R. Matthews, T. K. Maycock, T. Waterfield, Yelekçi, R. Yu and B. Zhou (eds.)]. Cambridge University Press. In Press. 2021.

2. United Nations Environment Programme and Climate and Clean Air Coalition (2021). Global Methane Assessment: Benefits and Costs of Mitigating Methane Emissions. Nairobi: United Nations Environment Programme.

3. Stolaroff, J.K., et al., Review of Methane Mitigation Technologies with Application to Rapid Release of Methane from the Arctic. Environmental Science andamp; Technology, 2012. 46(12): p. 6455-6469.

4. Sirigina, D.S.S.S., A. Goel, and S.M. Nazir. Multiple Greenhouse Gases Mitigation (MGM): Process Intensification To Mitigate Non-CO2 Gases And CO2 From Air. in AIChE Annual Meeting. 2021. Boston, USA.

5. Alyani, M. and K.J. Smith, Kinetic Analysis of the Inhibition of CH4 Oxidation by H2O on PdO/Al2O3 and CeO2/PdO/Al2O3 Catalysts. Industrial andamp; Engineering Chemistry Research, 2016. 55(30): p. 8309-8318.

6. Sabatino, F., et al., A comparative energy and costs assessment and optimization for direct air capture technologies. Joule, 2021.

628g Feasibility of CO_2 Direct Air Capture Integration in Data Centers

Lindsey Hamblin, Matthew D. Green, Klaus S. Lackner

Global carbon emissions have increased exponentially since the industrial revolution, leading to a rise in global temperature. To meet the 2°C warming limit put forth by the United Nations, the International Panel on Climate Change has stated that substantial increases in negative carbon emissions by the end of the century are now necessary[1]. Direct air capture of carbon dioxide is a promising negative emissions technology. However, due to the low atmospheric concentrations of CO_2 DAC technology is still energy-intensive and expensive[2]. Targeting niche applications where existing airflow and waste heat can be used advantageously may lead to the rapid development that can make larger DAC commercialization more economically viable.

DAC integration in data centers is one such application. There are over 2,700 data centers in the United States with an average power usage effectiveness of 1.43[3]. The best facilities can approach values closer to 1.09, where a PUE of 1 means all power consumption is used for processing power rather than cooling. These lower PUE values are produced by pushing ambient air through the servers or using evaporative cooling[4] rather than using traditional HVAC or cooling towers. Data centers currently use 200 TWh of electricity every year, 1% of the global energy demand[5]. This value is expected to increase to 7% of global demand by 2030. A substantial portion of that energy leaves data centers as thermal waste heat. This research aims to examine the feasibility of integrating data center airflow and waste heat into DAC systems.

Such a system must not interfere with the operation of servers and lead to a minimal increase in water and electricity usage. Therefore, the target is not to optimize the DAC process itself, but the operation of DAC within constraints placed by data center operation. This is made possible by using a series of staged vapor-liquid equilibriums to regenerate a liquid sorbent using only a 10°C differential that can be obtained from data center waste heat. Gas with a lower partial pressure of CO2 enters a reservoir and is adsorbed into a low-temperature liquid sorbent. This liquid sorbent then passes through a heat exchanger to a second hotter reservoir, where CO2 is released with a higher partial pressure than it first entered. This gas can be passed to the next stage until the desired CO2 concentration is achieved. ASPEN modeling is used to examine the energetics of using liquid sorbents in this staged "ladder" regeneration system and identify sorbents that minimize water loss in an air contactor. Results of promising sorbents are verified with experimental gas-liquid contactors and bench-scale prototypes of the regeneration system. This novel method of sorbent regeneration may provide an alternative for energy-intensive sorbent regeneration methods. Contactor design is optimized by testing a series of hollow fiber membranes in a lab-scale wind tunnel. The configuration with the lowest pressure drop and water loss will be ideal for DAC integration in data centers. Phosphate sorbents show promise in limiting water loss and reducing contact with more corrosive hydroxides in an expensive system.

[1] Masson-Delmotte, V., Zhai, P., Chen, Y., Goldfarb, L., Gomis, M. I., Matthews, J. B. R., Berger, S., Huang, M., Yelekçi, O., Yu, R., Zhou, B., Lonnoy, E., Maycock, T. K., Waterfield, T., Leitzell, K., andamp; Caud, N. (2021). *Working Group I Contribution to the Sixth Assessment Report of the Intergovernmental Panel on Climate Change Edited by*. www.ipcc.ch

[2] Holmes, G., andamp; Keith, D. W. (2012). An air-liquid contactor for large-scale capture of CO 2 from air. *Trans. R. Soc. A*, *370*, 4380–4403. https://doi.org/10.1098/rsta.2012.0137

[3] CloudScene, “Browse Markets,” n.d. (accessed April 2, 2021).

[4] *StatePoint Liquid Cooling system for data centers - Engineering at Meta.* (n.d.). Retrieved April 7, 2022, from https://engineering.fb.com/2018/06/05/data-center-engineering/statepoint-liquid-cooling/

[5]G Andrae, A. S., andamp; Edler, T. (2010). On Global Electricity Usage of Communication Technology: Trends to 2030. *Challenges*, *6*, 117–157. https://doi.org/10.3390/challe6010117

628h A Rooftop Approach for Direct Air Capture of Carbon-Dioxide

Shareq Mohd Nazir, Devesh Sathya Sri Sairam Sirigina, Mit Rakesh Surati, Shivani Ramprasad Jambur

The goal of max 1.5 °C of global warming by the end of the century cannot be reached without atmospheric CO_2 capture [1-3], and several approaches are being researched. Capacities for the following approaches until 2050 are (i) BECCS (0.5–5 $GtCO_2$/yr) (ii) afforestation and reforestation (0.5–3.6 $GtCO_2$/yr) (iii) DAC capture and storage – DACCS (0.5–5 $GtCO_2$/yr) (iv) enhanced weathering (2–4 $GtCO_2$/yr) (v) ocean fertilization (vi) biochar (0.5–2 $GtCO_2$/yr) (vii) and soil carbon sequestration (5 $GtCO_2$/yr) [3]. Each of these approaches has limitations. For example, reforestation, improved forest and crop management, and BECCS have land constraints, with a displacement of natural habitats and agriculture. Hence, DAC will be an essential component in the full range of Negative Emission Technologies (NETs), especially when addressing CO_2 emissions from sources that are not easy or cost effective to capture with BECCS. DAC has attributes of high value to the politics similar to that of emergency response [4]. It is inherently a cleanup approach. Such response benefits from (i) systems that are modular, scalable, and highly controllable by the society, government and industries/companies that invest (ii) CO_2 removals are verifiable (iii) deployment does not interfere with existing practices nor political/industrial interests. DAC does not have biophysical limits [5] unlike BECCS, nor does it require a large area as compared to reforestation. To capture 1 ton/yr CO_2, a unit for DAC requires a footprint of 0.4–24 m^2 (depending on the source of energy to provide regeneration heat), which can be compared to the forest footprint of about 900 m^2 to capture the same amount of CO_2. DAC can be considered an artificial and a quick method to remove CO_2 from atmosphere.

Among technologies for DAC of CO_2, wet scrubbing with aqueous alkali hydroxides [6] and vacuum temperature swing adsorption (VTSA) using supported sorbents [7] have gained attention. The wet-scrubbing technology requires a very high temperature (about 900 °C) in the regeneration of the solvent and also has challenges with the caustic recovery of alkali hydroxides [8]. The VTSA process with its aminated solids has an advantage because of lower temperatures of regeneration (80–130 °C) [9] required to regenerate the amines. The VTSA process also gives flexibility in using low-grade waste heat to regenerate the amines. In addition, the VTSA technology can be scaled from modular to large scale deployments.

In this study, we present system analysis for a DAC value chain with a modular rooftop approach. The DAC system is based upon a VTSA process interconnected to both the heat and electricity from renewable sources (mainly solar). Here, we consider DAC system to be installed on rooftops of household and commercial buildings supported by solar panels (thermal or PV) for both heat and electricity demands. The captured CO_2 is stored temporarily in small pressurized containers. The CO_2 is collected periodically from the buildings and sent to a main collecting terminal, from where it is transported for either utilization or permanent storage. This solution is similar to the smart-grid approach for renewable electricity. The advantages of such a value chain are (i) it optimizes the space and the costs for deploying DAC, since rooftops are generally freely available (ii) enables all the actors in the society to participate in CO_2 removal and thereby contribute to climate targets (iii) the energy penalty can be significantly reduced by capturing the CO_2 in the ventilation air (in addition to atmospheric air) which has higher concentration of CO_2 than in atmosphere. In this article, we present the potential of the envisioned modular rooftop DAC value chain in terms of (i) total solar energy required in top 10 countries with most CO_2 emissions (ii) total CO_2 removed in top 10 countries with highest solar energy potential. We also present the results in terms of cost per ton CO_2 captured, transported and utilized/stored.

Acknowledgements

This work is financially supported by the Swedish governmental initiative StandUp for Energy.

References

1. van Vuuren, D.P., et al., *Alternative pathways to the 1.5 °C target reduce the need for negative emission technologies.* Nature Climate Change, 2018. **8**(5): p. 391-397.
2. Minx, J.C., et al., *Negative emissions—Part 1: Research landscape and synthesis.* Environmental Research Letters, 2018. **13**(6): p. 063001.
3. Fuss, S., et al., *Negative emissions—Part 2: Costs, potentials and side effects.* Environmental Research Letters, 2018. **13**(6): p. 063002.
4. Hanna, R., et al., *Emergency deployment of direct air capture as a response to the climate crisis.* Nature Communications, 2021. **12**(1): p. 368.
5. Smith, P., et al., *Biophysical and economic limits to negative CO2 emissions.* Nature Climate Change, 2016. **6**(1): p. 42-50.
6. Keith, D.W., et al., *A Process for Capturing CO2 from the Atmosphere.* Joule, 2018. **2**(8): p. 1573-1594.
7. Wurzbacher, J.A., et al., *Concurrent Separation of CO2 and H2O from Air by a Temperature-Vacuum Swing Adsorption/Desorption Cycle.* Environmental Science andamp; Technology, 2012. **46**(16): p. 9191-9198.
8. Sabatino, F., et al., *Evaluation of a Direct Air Capture Process Combining Wet Scrubbing and Bipolar Membrane Electrodialysis.* Industrial andamp; Engineering Chemistry Research, 2020. **59**(15): p. 7007-7020.
9. Gebald, C., et al., *Stability of Amine-Functionalized Cellulose during Temperature-Vacuum-Swing Cycling for CO2 Capture from Air.* Environmental Science andamp; Technology, 2013. **47**(17): p. 10063-10070.

661a *Generalized Disjunctive Programming* (GDP) Model for the Optimal Capacity Planning of Reliable Power Generation Systems

Seolhee Cho, Ignacio Grossmann

As evidenced by the Texas power crisis in 2021, power generation systems should be designed to have high reliability to supply uninterrupted electric power to industries. Reliability measures the probability that a system will perform its required function accounting for the possible failures of equipment. One method to improve the reliability of the systems is to add redundant units (i.e., backup units) to prevent the system from completely failing. This approach is known as 'reliability-based design optimization,' and various studies on this topic have been conducted [1-3]. The authors of previous works mainly focus on optimizing the number of redundant units. However, since power systems operate in unsteady state due to time-varying power demand, the reliability is also influenced by the operational strategies that the systems use to satisfy the load demand. Specifically, the redundant units have a dual role: they can either remain as backups or else participate in power production depending on the power demand.

One of the conventional methods used to evaluate the reliability of power systems is "N-1 reliability". The N-1 reliability assumes that a power system can withstand an unexpected failure of a single component [4]. This implies that power systems may not function properly if multiple units fail simultaneously. The failures of multiple generators may reduce the power output but not necessarily fail the entire system. Hence, a rigorous method anticipating every possible failure state, and selecting the proper number and size of the backup generators should be developed to design and plan reliable power generation systems.

This paper aims to develop a new *Generalized Disjunctive Programming (GDP)* model for the rigorous optimal capacity planning of reliable power generation systems. The model optimizes the number and size of redundant units to maximize the reliability, and to minimize the cost by considering operation strategies that can affect the system reliability. We develop a GDP model, which involves Boolean and continuous variables, algebraic equations, and logic propositions, to represent the reliability and expected power production rigorously. It also includes in the objective function penalties for the loss of expected power. The resulting GDP model, which involves embedded disjunctions, can be reformulated as a mixed-integer linear programming (MILP) using either Big-M and hull-relaxation methods [5]. The proposed GDP model is tested with an expansion planning problem of a power generation system, and compared with a simple sequential design approach. The results show that the proposed model can effectively design reliable power generation systems, and yield significant economic savings compared to simplified design approaches.

References

[1] Chen, Y., Ye, Y., Grossmann, I.E., Chen, B., "Integrating Reliability and Uncertainty in Process Synthesis", *Computer Aided Chemical Engineering*, **50**, 107-114 (2021).

[2] Ortiz-Espinoza, A.P., Ye, Y., Grossmann, I.E., Jiménez-Gutiérrez, A., "Multi-Objective Optimization for the Incorporation of Safety and Reliability Considerations in Process Design", *Computer Aided Chemical Engineering*, **50**, 101-106 (2021).

[3] Ye, Y., Grossmann, I.E., Pinto, J.M., "Mixed-Integer Nonlinear Programming Models for Optimal Design of Reliable Chemical Plants", *Computers and Chemical Engineering*, **116**, 3-16 (2018).

[4] Ballireddy, T. R. R., andamp; Modi, P. K. Power System Expansion Planning Incorporating Renewable Energy Technologies with Reliability Consideration: A State of Art Literature. *International Journal of Recent Technology and Engineering,* **8**, 12403-12414 (2019).

[5] Grossmann, I.E., and Trespalacios, F., "Systematic Modeling of Discrete-Continuous Optimization Models through Generalized Disjunctive Programming," *AIChE Journal*, **59**, 3276-3295 (2013).

661b Lessons from an Integrated Optimisation Model of the UK Power, Heating, and Hydrogen Sectors

Matthias Mersch, Christos N. Markides, Niall Mac Dowell

Future energy systems are likely to be highly integrated. Increased electrification of heating and transport, and fuel switching from natural gas and petroleum products to blue or green hydrogen will likely be key enablers of the transition towards carbon neutral energy systems. Together with a higher share of volatile renewables in the power sector, with large-scale storage, firm generation capacity and demand-side management as balance mechanisms, this will break up the traditional structure of separated energy systems and markets.

Conventional energy system models tend to only consider single sectors, most commonly looking only at power generation. For future projections, they rely heavily on assumption regarding e.g., added demand from electrification of other sectors. Likewise, many studies on low-carbon heating technologies depend on electricity and/or hydrogen price assumptions. The enforcement of emission targets is also associated with implicit assumption. For example, it could be more economical to allow residual emissions from heating, which are then balanced by negative emissions from the power sector. Therefore, integrated energy system models of multiple sectors are of high interest when investigating energy system transitions.

We present here an integrated capacity-expansion and unit-commitment optimisation model of the electricity, heating, and hydrogen sectors in the UK. The development of technology portfolios is developed starting from today's installed capacities until 2050, allowing investment, retrofit and decommissioning decisions every 5 years. Additionally, the dispatch schedules are optimised on an hourly basis for each of the decision years, using either full-hourly time series or a typical-day clustering approach. The clustering method preserves periods of peak demand and low renewables availability, such as not to underestimate the required technology capacities. A central planner perspective is taken, meaning that all decisions are made by one entity with the target to minimise total system costs, rather than individual stakeholders minimising their costs or maximising their profits. Perfect foresight is assumed, which means that technology portfolios and operation are optimised over the entire time horizon simultaneously rather than applying a rolling horizon. The model is implemented as an LP model in GAMS, solved by Gurobi, with data pre- and post-processing done in Python.

The electricity sector is modelled using adapted equations from the original electricity sector model developed by Heuberger et al. [1][2]. A variety of electricity generation and storage technologies are considered in the model:

- Conventional thermal power plants (coal, open-cycle gas turbines (OCGT), combined-cycle gas turbines (CCGT), nuclear)
- Coal plants and CCGTs with carbon-capture and storage (CCS)
- Hydrogen-fired CCGTs and OCGTs
- Biomass plants with or without CCS (the former of which can provide negative emissions)
- Renewables (on- and offshore wind, solar, hydropower, geothermal plants)
- Interconnectors
- Large-scale storage (pumped hydro, Li-ion batteries)

Constraints include hourly demand matching, reserve and inertia provision, build-rate constraints to limit capacity expansion to reasonable levels and operational constraints of technologies. Additionally, capital, fuel and various other operational costs are calculated, as well as emissions.

The heating sector is split into three parts: domestic heating, commercial heating and industrial heating. Each come with their own demand, own set of technologies, and constraints. One important distinction to the power sector is that for the heating sector the demand has to be satisfied per end-user, since end-users cannot share heat with each other. For that reason, the domestic building stock is clustered into a number of representative buildings, and technology portfolios and operation are optimised for these representative buildings. The overall technology capacities, fuel use, costs and emissions are then aggregated based on the number of buildings that fall into each category. A similar approach is taken for commercial heating, where properties are split into office and retail buildings, restaurants and cafes, and warehouses/storage facilities/industrial buildings. In each category, a high-demand and a low-demand property is considered. Industrial heating is split into high-temperature and low-temperature demand. Technologies considered for heating are natural gas and hydrogen boilers, air-source and ground-source heat pumps and electric resistive heating, as well as water tanks for storage.

Hydrogen is an option as fuel for both, low-carbon heating and electricity. It can also function as storage medium. Following a similar structure as the electricity sector, hydrogen generation and storage capacities are optimised, and ideal dispatch schedules are determined. Considered generation technologies are methane reformers with CCS, electrolysers and biomass gasification with CCS, which can provide negative emissions, similar to biomass power generation with CCS. Salt caverns and line pack are considered as storage options.

The different sectors are explicitly integrated in the modelling. Electrification of heating for example will result in a higher electricity demand, which in turn requires more electricity generation assets. The same is true for the hydrogen sector, where all demand is due to heating or electricity generation, no exogenous hydrogen demand is assumed. This integration and simultaneous optimisation eliminates many of the aforementioned key uncertainties related to single-sector energy system models.

We present results from different optimised energy system transition pathways, focussing on trade-offs and interactions between the different sectors. The integrated model allows us to explicitly quantify the impact of decisions in one sector on the others. Additionally, we are able to determine optimal emission trajectories for the different sectors that lead net-zero overall emissions. The prioritisation of resource use, e.g., is it better to use biomass for power generation or for hydrogen generation via gasification, can also be evaluated using the integrated model.

References

[1] C. F. Heuberger, I. Staffell, N. Shah, and N. Mac Dowell, "A systems approach to quantifying the value of power generation and energy storage technologies in future electricity networks," *Comput. Chem. Eng.*, vol. 107, pp. 247–256, Dec. 2017, doi: 10.1016/j.compchemeng.2017.05.012.

[2] C. F. Heuberger, E. S. Rubin, I. Staffell, N. Shah, and N. Mac Dowell, "Power capacity expansion planning considering endogenous technology cost learning," *Appl. Energy*, vol. 204, no. August, pp. 831–845, 2017, doi: 10.1016/j.apenergy.2017.07.075.

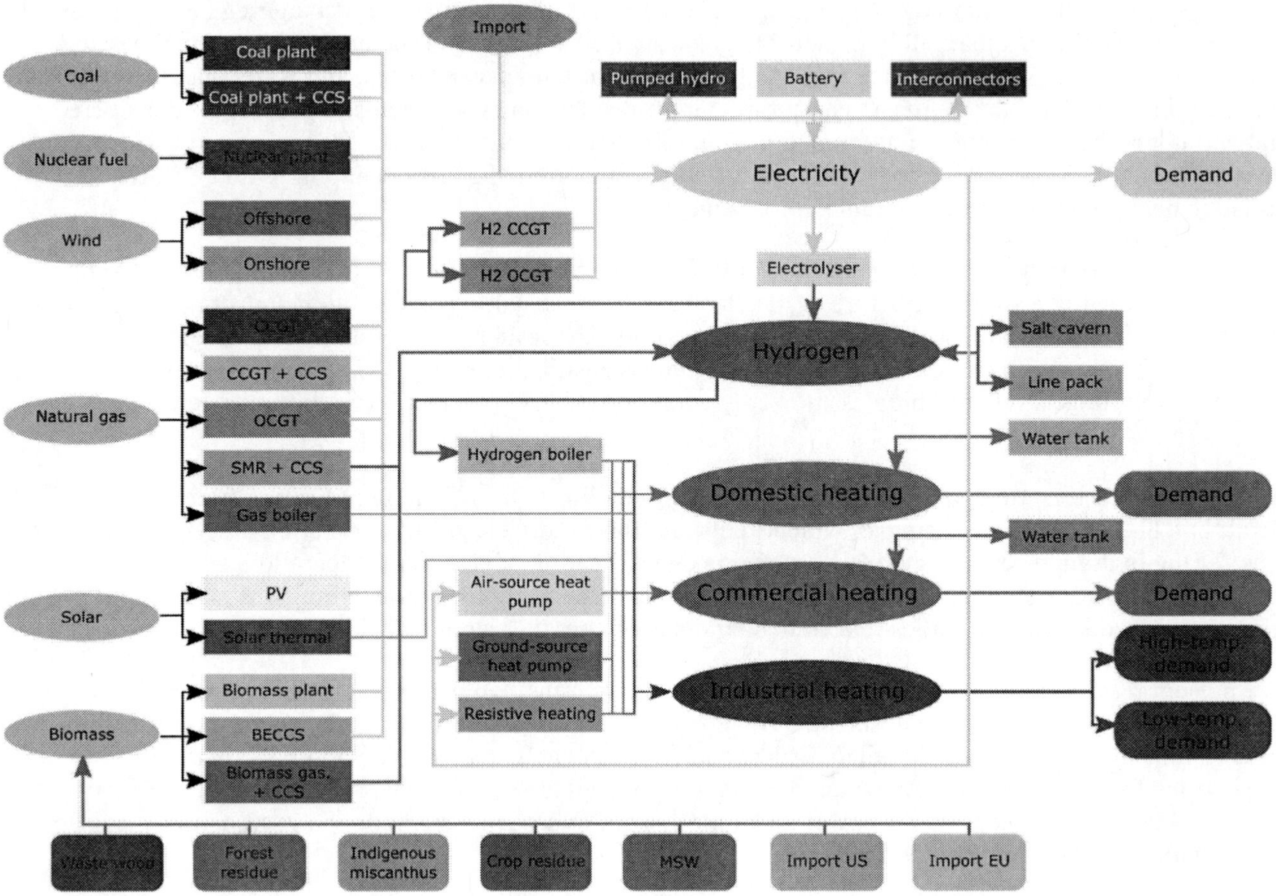
Import
Pumped hydro
Battery
Interconnectors
Coal
Coal plant
Coal plant + CCS
Nuclear fuel
Nuclear plant
Electricity
Demand
Wind
Offshore
Onshore
H2 CCGT
H2 OCGT
Electrolyser
CCGT
Hydrogen
Salt cavern
CCGT + CCS
Line pack
Natural gas
OCGT
Water tank
Hydrogen boiler
SMR + CCS
Domestic heating
Demand
Gas boiler
Water tank
Air-source heat pump
PV
Commercial heating
Demand
Solar
Solar thermal
Ground-source heat pump
High-temp demand
Biomass plant
Industrial heating
Resistive heating
Biomass
BECCS
Low-temp demand
Biomass gas. + CCS
Waste wood
Forest residue
Indigenous miscanthus
Crop residue
MSW
Import US
Import EU

661c GHG Emissions Reduction By Optimizing Design and Operation of Cross-Sector Integrated Energy Systems: Civic and Industrial Sectors

Ruonan Li, Vladimir Mahalec

Individually operating combined cooling, heating, and power (CCHP) system has been shown can decrease greenhouse gas (GHG) emissions, increase energy efficiency, and improve the flexibility of the power system [1] by generating heating, cooling, and electricity locally at the same time. However, there remains an open question on whether the GHG emissions can be further reduced by integrating the individual CCHP systems across residential, commercial, industrial, and transportation sectors through heating and electricity transfer.

Typically, a CCHP system operates either following the electricity load (FEL) or following the thermal load (FTL). The power generation unit (PGU) of the CCHP system generates electricity and heat by combusting natural gas, where the heat is recovered for meeting heating and cooling demands. However, the heat and electricity generated usually cannot be entirely consumed at the same time, which leads to excess or shortage in either the heat or electricity. The mismatch between the energy generation and supply can lead to the system operating supplementary energy system equipment (such as the backup boiler) or using supplementary energy resources (such as the grid electricity), which leads to higher GHG emissions and total costs.

The unbalanced heat and electricity demand and supply can also be solved by integrating multiple CCHP systems by transferring electricity and heat among entities. Each entity in the network performs both as a supplier and a consumer to dispatch excess energy to entities that face energy shortages or to receive energy from other entities when energy demands cannot be fully satisfied. To extensively balance the energy demands and supplies of the overall integrated system, entities with different energy demand patterns should be combined. For example, an integrated system can combine residential buildings and commercial buildings that have higher electricity demand than heating demand with food processing plants that require more heating than electricity, such as confectionery plants, breweries, and bakery plants. Thus, the residential and commercial buildings can transfer excess heat to the plants, while receiving surplus electricity from the plants.

Several studies have been performed on the integration of energy systems, for example, studies performed by Mehleri et al. [2], Li et al. [3], and Jing et al. [4]. However, these studies: 1. do not consider the temperature of heating demands when modeling the heat transfer; 2. do not consider opportunities for industrial plants to switch their production volumes, thus, adjusting their heat and electricity usages based on the energy demands of other entities; 3. have not explored the relationship between the entity sizes and maximum GHG emissions reduction of the integrated operation.

To address the research gaps mentioned above, this work investigates the design and operation of the integrated systems for residential, commercial, and industrial sectors through optimization approaches. Besides capacity and the amount of heating, cooling, or electricity generated by each piece of the energy system equipment, the approach also finds the optimal industrial production volumes. The industrial processes that require heating have been differentiated as the ones that can use the transferred heat from other entities (processes at low temperatures) and the ones that cannot use the transferred heat (processes at high temperatures). Heat balances for each of the industrial processes have been developed, reflecting whether the processes can use the transferred heat to ensure feasible heat transfer among entities.

The optimal solution identifies the operation and capacity of the energy system equipment, as well as production volumes of the plants that minimize GHG emissions and annual total costs (capital costs and operation costs) of the entire integrated system. Since there is a trade-off between GHG emissions and annual total costs (ATC) of the CCHP system, Pareto fronts of the integrated system are employed to depict trade-offs between non-dominated solutions.

In this work, an integrated system with a residential building (including electric vehicles), a supermarket, a brewery, a confectionery plant, and a bakery plant has been used for case studies. Except for the supermarket, each of the entities has been assumed to have energy system equipment, as shown in Figure 1. Thus, besides natural gas, the system also utilizes solar energy to generate heating, cooling, and electricity. Since the supermarket has extremely high electricity demand for providing cooling at low temperatures, using an absorption chiller and PGU in the supermarket is not efficient. Thus, the supermarket has been assumed does not implement the PGU, absorption chiller, and heat recovery unit. The integrated system has been compared to a non-integrated system. The non-integrated system is identical to the integrated system; however, there is no energy transfer among entities.

Results of the case studies show that, under the same ATC values, the integrated system releases over 18.8% less GHG emissions. The lower GHG emissions are because, with energy transfer, operations of the PGUs in the plants are no longer limited by electricity demands (lower compared to heating demands) of the plants. The confectionery plant, the brewery, and the bakery plant operate PGUs more frequently to generate more heating used in the entities and transfer the associated excess electricity to the residential building, which has higher electricity demand. As a result, the integrated system uses 93% less natural gas for operating backup boilers and purchases over 66% less electricity from the grid compared to the non-integrated system, which further leads to lower GHG emissions.

By optimizing production volumes, GHG emissions reduction of the integrated system (compared to the non-integrated system) can be further increased by 2% compared to the integrated system with fixed industrial production volumes at average daily rates. Additionally, it has been found that when temperatures of heating processes are not differentiated, around 34% of the heat transfer is infeasible, where heat at lower temperatures is transferred to processes that require high-temperature heating.

The relationship between the sizes of entities and GHG emissions reduction achieved by the integrated system has been explored by adapting the objective function of the optimization problem and setting the sizes of entities as decision variables. The investigation intends to find whether there exists an optimal relative entity size that maximizes GHG emissions reduction of the integrated system compared to the non-integrated system. Therefore, the objective function has been changed to maximize the GHG emissions reduction percentage between the integrated system and the non-integrated system, as shown in Eq. (1). The objective function is based on the minimum GHG emissions of the integrated system and the non-integrated system, where the minimum GHG emissions for the non-integrated system can be expressed as a linear equation of entity sizes.

Case studies have been performed on the same system mentioned above, except that the entities no longer implement batteries for storing electricity. The results show that the integrated operation can reduce GHG emissions by a maximum of 17%, compared to the non-integrated system. It requires the system to combine the residential building and the supermarket with 870 electric vehicles, a bakery plant with a production volume of 5.0×10^3 kg/day, and a brewery that generates 2.7×10^3 kg product each hour. The ratio between sizes of the residential building, supermarket, electric vehicles, bakery plant, and brewery represents the optimal relative entity size, following which GHG emissions reduction benefits of the integrated operation can be maximized.

When there are requirements on sizes of specific entities that result in the optimal relative entity sizes cannot being followed, GHG emissions reduction benefits provided by the integrated operation can be lower. However, as shown in Figure 2, by optimizing the sizes of other entities, the reduction can still be maintained at a value close to 17%.

The studies show that by integrating individually operating CCHP systems, GHG emissions of the overall system can be further reduced (by 17% for the system in case studies) compared to the corresponding non-integrated system. The maximum GHG emissions reduction benefits require the integrated system to be built following the optimal relative entity sizes, optimize production volumes of industrial plants, and optimize the operation and capacity of energy system equipment. When there are requirements on the sizes of specific entities in the integrated system, by optimizing the sizes of other entities, a GHG emissions reduction close to the maximum value can still be achieved. Additionally, it is necessary to differentiate the temperature of heating demands to ensure feasible heat transfer among entities. The optimization approaches introduced make a further step in reducing GHG emissions of distributed energy systems through integrated operation.

References

[1] Mancarella P, Chicco G, Capuder T. Arbitrage opportunities for distributed multi-energy systems in providing power system ancillary services. Energy 2018;161:381–95. https://doi.org/10.1016/j.energy.2018.07.111.

[2] Mehleri ED, Sarimveis H, Markatos NC, Papageorgiou LG. A mathematical programming approach for optimal design of distributed energy systems at the neighbourhood level. Energy 2012;44:96–104. https://doi.org/10.1016/j.energy.2012.02.009.

[3] Li L, Mu H, Gao W, Li M. Optimization and analysis of CCHP system based on energy loads coupling of residential and office buildings. Appl Energy 2014;136:206–16. https://doi.org/10.1016/j.apenergy.2014.09.020.

[4] Jing R, Wang M, Liang H, Wang X, Li N, Shah N, et al. Multi-objective optimization of a neighborhood-level urban energy network: Considering Game-theory inspired multi-benefit allocation constraints. Appl Energy 2018;231:534–48. https://doi.org/10.1016/j.apenergy.2018.09.151.

$$max \frac{GHG_{non-integrated} - GHG_{integrated}}{GHG_{non-integrated}} \quad (1)$$

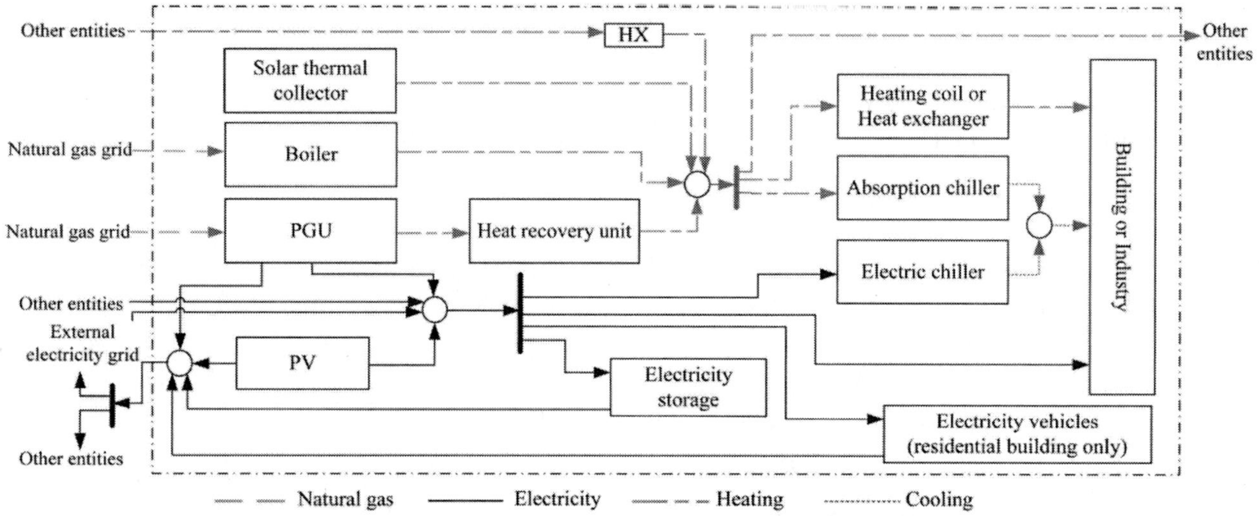

Figure 1. Representation of the energy system of an entity.

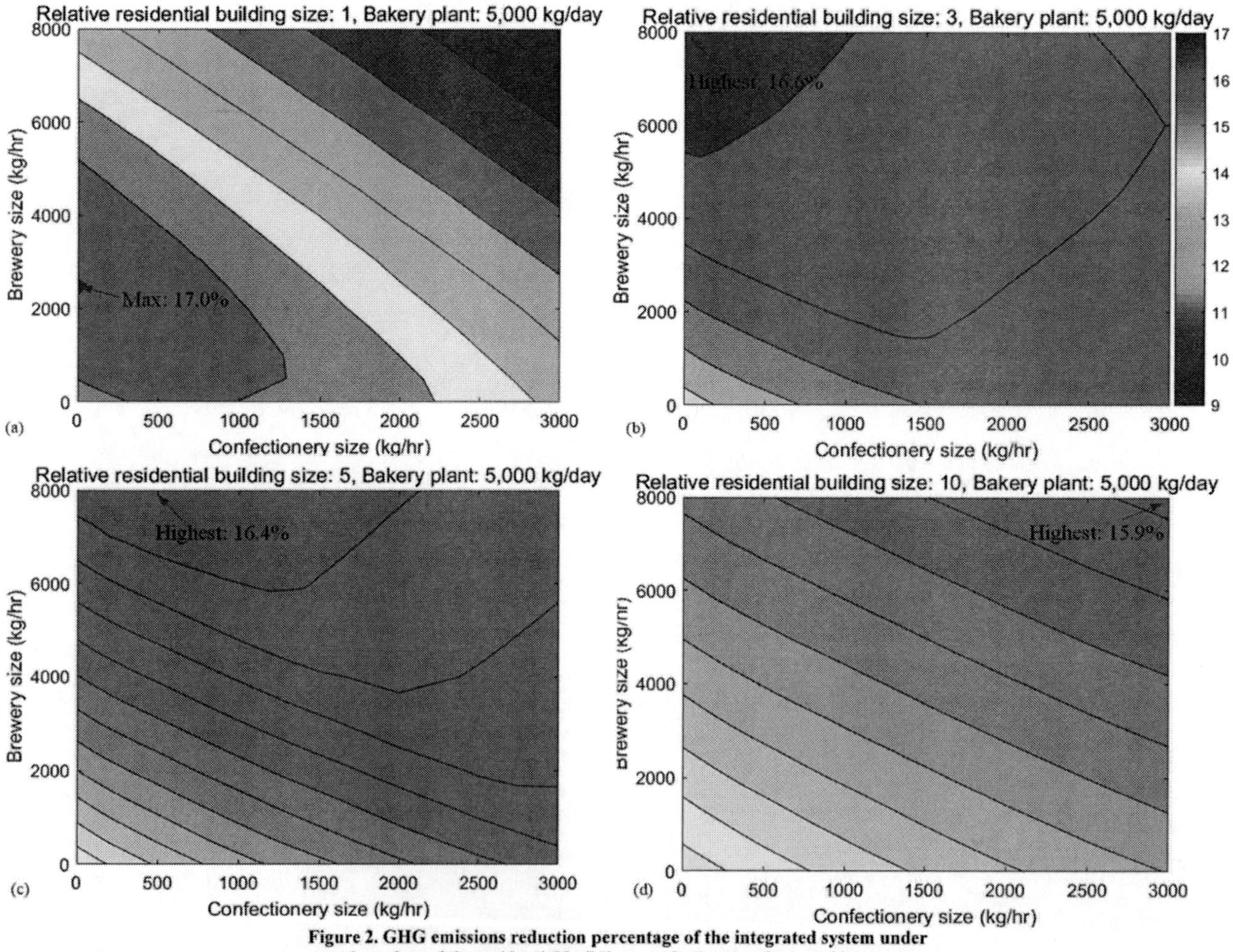

Figure 2. GHG emissions reduction percentage of the integrated system under various sizes of the residential building, confectionery plant, and brewery.

Long-term expansion planning of an integrated power and energy carrier generation system: A tailored decomposition algorithm

Ioannou I., Eidgenössische Technische Hochschule Zürich, Zurich, Switzerland
Dr. Grossmann I., Carnegie Mellon University, Pittsburgh, PA, United States
Dr. Guillén Gosálbez G., Eidgenössische Technische Hochschule Zürich, Zurich, Switzerland

Abstract

Integrating a power system's generation and transmission expansion planning (GTEP) with the management of an energy carrier supply chain (ECSC) provides an appealing framework to reduce power curtailment when the supply exceeds the load demand. This, in turn, may reduce the system's storage and power transmission requirements. Standard formulations for the GTEP of power systems involve large-scale mixed-integer linear programming (MILP) models with several millions of variables (Li et al., 2022a). Integrating an ECSC into the GTEP model further increases computational complexity. Several methods have received significant attention to circumvent the latter challenges based on relaxation, and decomposition strategies, such as block structure decomposition (Benders, 1962), and rolling horizon (RH) (Zamarripa et al., 2016), among others. Furthermore, nested decomposition strategies can further accelerate the solution process (Lara et al., 2018). This study proposes combining Benders decomposition (BD) with a RH approach to address the challenge of reducing the computational effort. We evaluate the performance of the suggested algorithm on an MILP of a GTEP-ECSC system. The computational experiments show that the proposed scheme can reduce CPU time by almost 50-70% compared to the stand-alone BD algorithm, with only minor optimality deviations.

Introduction

Electricity is at the heart of our modern world, with benefits which include ending poverty, supporting human health and education, reducing inequalities between countries, and enhancing economic growth. Unfortunately, the expected rising load demand may have detrimental effects on the environment since electricity production is also responsible for releasing large amounts of greenhouse gases (GHGs). With the global aim to significantly reduce GHG emissions aligned with the Paris climate agreement (United Nations, 2016), the power sector could play a pivotal role. Besides, the production and use of energy across economic sectors contribute 75% of the EU's total GHG emissions (European Commission, 2019).

For that purpose, policymakers discuss adopting several existing and emerging technologies to satisfy the rising electricity demand in a more environmentally friendly way. Besides, the International Energy Agency scenarios show significant future reliance on solar and wind power (IEA, 2019). At the same time, thermal generators' carbon capture and storage (CCS) is considered a promising solution to reduce CO_2 emissions at their source (Bui et al., 2018). Furthermore, renewable electrolytic hydrogen (eH_2), either stand-alone or combined with carbon capture and utilization (CCU) or ammonia production, is another highly discussed transition (Demirhan et al., 2020). Nonetheless, the asynchronous nature of fluctuating renewables calls for careful investment and operation for energy storage and, thus, the reduction of output curtailment (Ioannou et al., 2022). Furthermore, expanding towards fluctuating power sources can, in turn, make it challenging to satisfy the demand or operate correctly under the worst plausible conditions (Cho et al., 2022; Li et al., 2022b). Finally, regions with high availability of renewables may not match demand locations, and thus, operating the power transmission effectively or ECSC might be crucial (Demirhan et al., 2020).

Within this spirit, an effective re-configuration of the power sector stresses the generation and transmission expansion planning (GEP and TEP, respectively, or their combination, GTEP), which, at present, has attracted significant interest (Lara et al., 2018; Li et al., 2022a). Such expansion planning aims to determine the location, size and type of power generating units (or transmission lines) for satisfying the long-term demand-supply. Namely, the GEP systems focus on power generation, while the TEP system monitors exchanges among producers and consumers via long-distance and high-voltage transmission lines (Conejo et al., 2016).

As mentioned above, there is an opportunity of incorporating EC(s) in the power system to tackle the stochastic nature of renewables. This, in turn, leads to sector-coupled supply chain management that goes beyond the concept of a typical profit-oriented enterprise (Flint, 2004). Besides, both systems could partly decarbonize their operations by embracing reciprocal connections (Allen et al., 2022; Demirhan et al., 2020). Finally, due to the vast energy requirements and efficiency losses for producing ECs (Ioannou et al., 2021), technology learning curves and the use of otherwise curtailed power could significantly influence the economic viability of such supply chains. Thus, the ECSC system merits integration with the power system's GTEP.

GEP and TEP models are usually large-scale MILP problems with yearly investment-related decisions. At the same time, operating decisions require an hourly representation within the long term, which increases the model's complexity significantly, thus, making it computationally intractable. The GTEP-ECSC integration further increases the system's complexity since they have inherently interconnected decisions. Besides, such a model should capture: (i) the sub-regional hourly variability of renewables, load demand, and ramp-up/ramp-down of thermal generators, (ii) the inter-regional hourly power transmission, (iii) the inter-regional hourly ramp-up/ramp-down of the EC production, (iv) the inter-regional daily EC transport, and (v) the annual investment decisions.

Problem statement

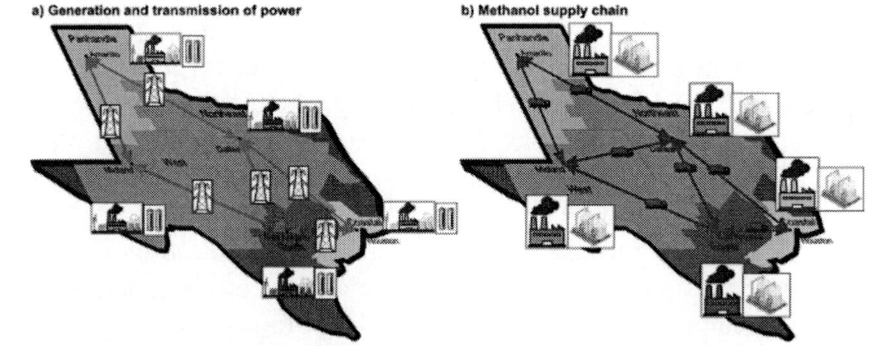

Figure 1: Representation of the system.

We aim to conduct computational experiments for a tailored Benders decomposition horizon (TBDH) algorithm on a large-scale GTEP-ECSC system. The model's objective function (OF) is the minimum net present cost for the related investment and operation activities from 2020 until 2050. We consider a methanol supply chain as a reference ECSC. Since most methanol plants in the United States are located in the Gulf Coast region (eia, 2019), we consider Texas the geographic area for our analysis. At the same time, Texas is an appropriate representation of future GTEP systems due to favourable renewable potential. The spatial space is discretized in five regions, and we define six possible busses connections and seven vehicle routes between them, as depicted in **Figure 1**. Hence, transport and transmission networks ensure the availability of methanol and power at the demand locations.

Given are the years $n \in N$, days $d \in D$, and hours $h \in H$ of the time horizon. Given are also the locations $l \in L$, utility-battery systems $b \in B$, types of power generation units $u \in U$, and transmission lines connections $t \in T$ for the designated geographic area. The D representative days are based on vector averaged regional data that are aggregated via hierarchical agglomerate clustering to simplify the temporal representation and, thus, decrease the computational effort. Furthermore, we consider $m \in M$ methanol production technologies, $p \in P$ precursors production technologies, $v \in V$ vehicles, and storage warehouses in the demand locations. Namely, the power generation units, u, consist of natural gas and coal power plants with and without CCS, wind onshore, solar photovoltaics, and nuclear and biomass plants. Regarding the methanol technologies, m, we consider production from natural gas or CO_2 hydrogenation. Therefore, we also consider technologies for eH_2 and CO_2. We consider existing and newly developed facilities (either for power or materials production). Moreover, trucks transport methanol while omitting the transportation of CO_2 and eH_2, i.e., produced in the demand locations. Finally, the technical and economic data, along with their direct CO_2 emissions, are given for installing and operating (i) power generators and storage technologies (NREL, 2021), (ii) electrolyzers (IEA, 2020), (iii) direct air capture (DAC) units (Fasihi et al., 2019), (iv) methanol production and storage (González-Garay et al., 2019; Towler and Sinnott, 2012), and (v) transport units (Kostin et al., 2011).

The model is multi-period one, where we consider the time evolution (i) of the power load (ERCOT, 2016) and methanol demand (eia, 2019) for the system regions (with an annual growth rate), (ii) for fuel costs (eia, 2019), and (iii) for technology learning curves (Fasihi et al., 2019; IEA, 2020; NREL, 2021). Furthermore, we consider the individual hourly solar irradiation and wind speed for the given region (Blair et al., 2014; Draxl et al., 2015; Sengupta et al., 2018). In addition, we set a linearly increasing taxation of direct CO_2 emissions (0–350 \$ t^{-1} for 2020 and 2050, respectively), along with quotas for the shares of renewables in the power system that act as lower bounds (0–50% for 2020 and 2050, respectively, with a linear increase). Finally, a penalty is given for the annual renewable-based power curtailment.

Due to space limitations, we only provide a concise representation of the problem and a brief description. The multiperiod MILP model is as follows:

$$OBJ = \min_{x,y} \sum_{n \in N} \sum_{l \in L} \left(c_n^{\top} x_{n,l} + \sum_{d \in D} W_d f_n^{\top} y_{n,l,d} \right) \qquad (1)$$

$s.t.$

$$A_n x_{n,l} + B_n y_{n,l,d} \leq b_{n,l,d} \quad \forall n \in N, d \in D, l \in L \quad (1a)$$

$$C_{n-1} x_{n-1,l} + E_n x_{n,l} \leq g_{n,l} \quad n = 2,3,\ldots,|N|, l \in L \quad (1b)$$

$$x_{n,l} \in X, \ \forall n \in N, l \in L, \quad (1c)$$

$$y_{n,l,d} \in Y, \ \forall n \in N, d \in D, l \in L \quad (1d)$$

where the variable $x_{n,l}$ represents investment decisions at year n and location l (including both discrete and continuous variables), $y_{n,l,d}$ represents operating decisions corresponding to the representative day d in year n and location l (including both discrete and continuous variables), while W_d is the weight of the representative day d to make the operating and investment costs comparable. Eq. (**1*a***) describes the operational decisions of each year n and each day d and location l, i.e., generation, transmission, power storage, and the unit commitment, along with the production, transportation, and storage of methanol. Eq. (**1*b***) are investment-related constraints, such as installing power and material production technologies, transmission lines between regions, purchasing utility batteries and vehicles, and expanding capacity, where we consider both the variable and fixed cost terms. At the same time, as the demand dictates, the model follows a dynamic annual expansion approach, where power and chemical production technologies are installed. Finally, Eq. (**1*b***) also describes the units' retirement or the operating life extension, if possible, of units at the end of life.

We solved the model in GAMS 38.2, with (i) the CPLEX solver, in non-parallel and parallel mode, (ii) the BD implementation from CPLEX (Bonami et al., 2020) in parallel mode, and (iii) with 3-scenarios, and with two different representative days, for the TBDH algorithm in parallel mode. We used an Intel CoreTM i7- 8700 machine at 3.20 GHz and 32 GB RAM. Finally, since the implementation of standard BD requires the relaxation of the integrality constraints of the $y_{n,l,d}$ operating variables, i.e., generators that are on, starting up, and shutting down, we can only obtain a LB to the original problem.

Solution Strategy and Results

CPLEX solver

Table 1 provides the statistics of solving the MILP (with relaxed $y_{n,l,d}$) using CPLEX. With only 2 or 3 representative days, non-parallel execution requires 4.6 and 16.7 h of CPU time, respectively. Parallel programming reduces the latter CPU times significantly. However, obtaining a solution with 4 representative days in less than 24 h of computation is impossible.

Table 1. CPLEX solver (with relaxed $y_{n,l,d}$)

Representative days (#)	**2**	**3**	**4**
Equations (Million)	1.23	1.84	2.45
Cont. var. (Million)	0.92	1.37	1.83
Disc. var. (Thousand)	14	14	14
OF | non-par. (B$)	309.7	323.6	–
Opt. gap | non-par. (%)	0.43	0.45	–
CPU time | non-par. (h)	4.6	16.7	24.0
OF | parallel (B$)	311.2	322.9	–
Opt. gap | par. (%)	0.96	0.27	–
CPU time | par. (h)	1.8	6.9	24.0

Standard Benders decomposition

BD was first introduced to solve MILP problems and, in short, divides the original formulation into a master problem (MP) and one or several sub-problems (SPs) (Benders, 1962). A generalized BD approach was later introduced to tackle MINLP (Geoffrion, 1972). The solution strategy involves the iterative introduction of cuts to the MP obtained from the SPs, narrowing the search space until convergence is achieved. By fixing the complicating variables of the MP, the SPs reduce to ordinary linear programs. In the GTEP-ECSC model, the MP includes the investment decisions, while the SPs consider the operating decisions. Finally, due to the block structure, each SP represents a year n, and thus, satisfies the demand independently.

Table 1. Standard BD solved in parallel ($y_{n,l,d}$ relaxed)

Representative days (#)	**2**	**3**	**4**	**5**	**6**	**7**	**8**	**9**	**10**	**11**
Equations (Million)	1.2	1.8	2.5	3.1	3.7	4.3	4.9	5.5	6.1	6.7
Cont. var. (Million)	0.9	1.4	1.8	2.3	2.7	3.2	3.6	4.1	4.6	5.0
Disc. var. (Thousand)	14	14	14	14	14	14	14	14	14	14
OF value (B$)	310.3	324.0	325.1	338.4	350.8	353.2	353.3	354.0	355.0	355.1
Opt. gap (%)	0.71	0.83	1.00	0.81	0.79	0.96	0.78	0.97	0.69	0.8
CPU time (h)	0.4	1.4	1.7	2.6	3.9	4.2	5.2	6.8	10.2	12.6

Table 2 shows that the BD allows increasing the number of representative days (i.e., temporal resolution) significantly while providing nearly optimal solutions (gap≤1.0%). Besides, the BD meets the stopping criterion with a 4.5- and 4.9-fold lower CPU time than CPLEX for 2 and 3 representative days, respectively. Here, we did not investigate solutions for more than 11 representative days, which requires 12.6 h of CPU time to meet the stopping criterion, since the OF value does not change significantly. **Tables 1-2** also illustrate that the main bottleneck of the proposed GTEP-ECSC model is the continuous variables and not the integer, which increase substantially as the number of representative days increases. Notably, this occurs since we relaxed the integrality condition of the $y_{n,l,d}$ operating variables of the thermal generators.

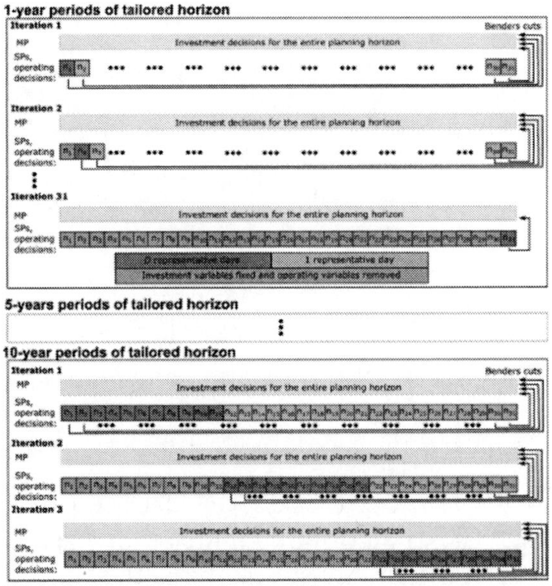

Figure 2: Overview of three scenarios with the TBDH algorithm.

Tailored Benders decomposition horizon (TBDH) algorithm

The two pillars of the suggested TBDH algorithm are the BD and the RH method that classifies between horizon blocks (i.e., detailed and aggregated ones).

Some works applied the RH method by relaxing the integrality condition for a subset of the original problem variables, i.e., the aggregated block. Following an iterative procedure, a set of discrete variables is fixed, and then, a new subset is investigated until the entire horizon is explored (Kostin et al., 2011; Zamarripa et al., 2016). Another alternative is to consider a simplified model for the aggregated block, compared to the full-space one, while following the same iterative concept as before (Zamarripa et al., 2016). Finally, one can divide the original formulation into a set of SPs with increasing horizon lengths that are solved iteratively until the entire horizon is covered. The temporal representation grows successively in each iteration, while a subset of binary variables is fixed (Liu et al., 2010).

Here, we aim to combine the BD with the RH while keeping the integrality condition of the investment decision and the original formulation unchanged. We do this by using the same MP as in the standard BD while capitalizing on normal- and reduced-sized horizons, i.e., set of SPs (**Figure 2**). The normal-sized SP(s) horizon considers a large number of representative days. In contrast, the reduced-sized horizon has only one representative day to represent each SP set. We solve the horizon iteratively, where we fix the investment decisions for the normal-sized SPs of the current iteration. Subsequently, a new SP set obtains a higher resolution. Furthermore, due to the block structure of BD, the operating variables are independent between SPs. Thus, in every iteration, the algorithm discards several hundred thousand continuous variables, thus, reducing in size, which is illustrated later. Units' retirement or life extension might occur during a detailed horizon, where the investment decisions are fixed from previously calculated horizons. This procedure continues until we cover the entire horizon.

We execute our computational experiments twice to investigate the effectiveness of the proposed algorithm with different time resolutions. Namely, these cases consider normal-sized SP(s) with 6 or 11 representative days. For both time resolutions, we consider three scenarios, as illustrated in **Figure 2**. Each system has 31 SPs that generate the BD cuts, i.e., equal to the cardinality of set n. In the first scenario, we consider only a single normal-sized SP in the detailed horizon, leading to 31 iterations of

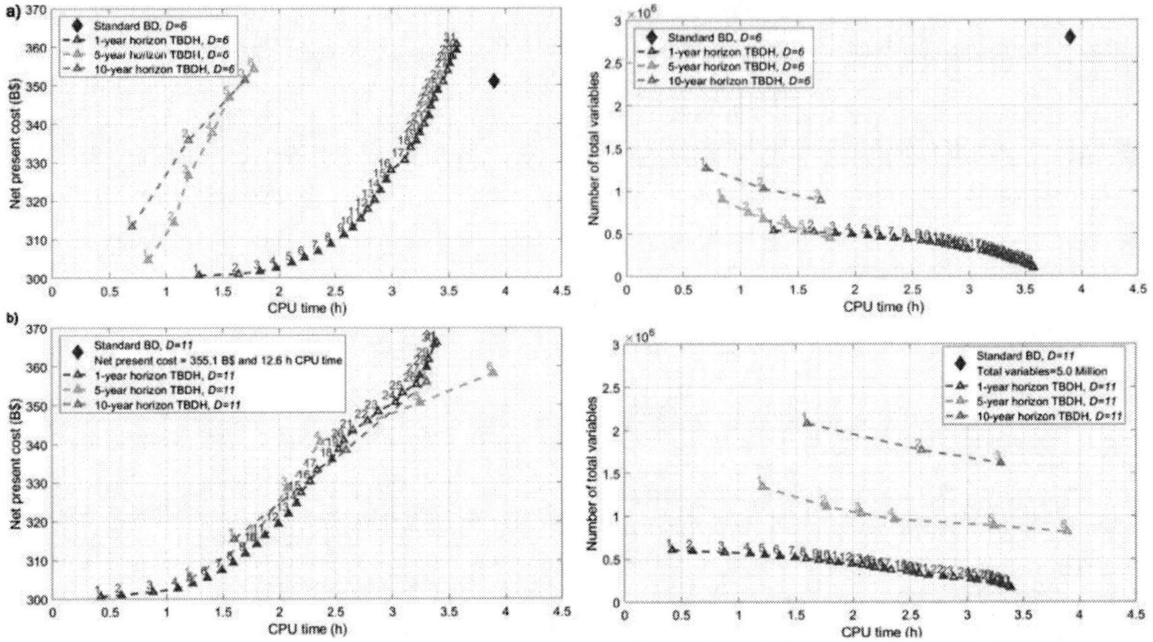

Figure 3: Objective function value (left) and total number of variables (right) vs. CPU time to obtain the optimality gap criterion for the three scenarios solve with the TBDH algorithm with (a) 6 (top) and (b) 11 representative days (bottom) and compared to the standard BD (**Table 2**). The numbers represent the iteration of the TBHD algorithm.

the TBDH algorithm. In the second scenario, we consider a 5-SPs normal-sized horizon, thus, requiring 6 TBDH iterations. Finally, in the third scenario, we use a 10-SPs normal-sized horizon, requiring only 3 TBDH iterations.

Figure 3 (left) shows that TBDH decreases the CPU time by almost {2.2–2.3}- and {3.2–3.9}-fold compared to the stand-alone BD for the {5 or 10}-SPs in the detail horizon and when considering 6 and 11 representative days, respectively. At the same time, the TBDH obtains an OF value only slightly different to that of BD (by {1.0–0.2}% and {0.9–0.2}% for the {5 or 10}-SPs horizon and for 6 and 11 representative days, respectively. The latter CPU time benefits, along with the minor deviations from optimality, from the tight approximation of the GTEP-ECSC. Besides, the aggregated horizon uses an average value for the future hourly demand and wind and solar generators' output. Notably, the single SP detail horizon does not lead to significant savings in CPU time for the 6 representative days resolution (~10%), while estimating a 2.7% higher OF value compared to standard BD due to error accumulation. In contrast, when considering 11 representative days and a single SP in the detail horizon, we observe significant CPU time benefits compared to BD (3.7-fold faster) while still having a high deviation from the OF value (by +3.1%). This occurs since the higher temporal resolution (D=11) does not increase significantly the continuous variables, which is not the case in the enlarged horizon counterparts that consist of many SPs (**Figure 4**, right). The latter effect is milder when considering 6 representative days.

Conclusions

In this paper we analyzed different solution strategies for an integrated GTEP-ECSC MILP model. The use of a commercial solver restricted the increase of the temporal representation to more than 3 representative days and a 24 h CPU time budget. The standard BD allowed solutions with a small optimality gap (gap≤1.0%) with significantly lower CPU time (≤13 h). BD could also yield solutions even when the number of representative days increased significantly (≤11). Combining the BD with the RH algorithm can reduce even further the computational effort compared to the BD (by ~50-70%). This algorithm capitalized on normal- and reduced-size SPs horizons based on the number of representative days. At the same time, the algorithm retains the integrality conditions of investment decisions and the original formulation. The deviation from the standard BD is minor when considering an enlarged SPs set in the detailed horizon. In contrast, a single SP detailed horizon led to a higher, but still small, deviation from the BD solution. At the same time, the CPU benefits of the TBDH algorithm are reduced when we consider a single SP detail horizon and a large average number of representative days ($D = 6$). However, when increasing the number of representative days ($D = 11$) the CPU benefits of a single SP detailed horizon increase drastically and become comparable to that of the enlarged horizon counterparts.

Acknowledgements

This publication was created as part of NCCR Catalysis (grant 180544), a National Centre of Competence in Research funded by the Swiss National Science Foundation.

References

Allen, R.C., Baratsas, S.G., Kakodkar, R., Avraamidou, S., Demirhan, C.D., Heuberger-Austin, C.F., Klokkenburg, M., Pistikopoulos, E.N., 2022. A multi-period integrated planning and scheduling approach for developing energy systems. Optim. Control Appl. Methods.

Benders, J.F., 1962. Partitioning procedures for solving mixed-variables programming problems. Numer. Math. 4, 238–252.

Blair, N., Dobos, A.P., Freeman, J., Neises, T., Wagner, M., Ferguson, T., Gilman, P., Janzou, S., 2014. System Advisor Model, SAM 2014.1.14: General Description. United States.

Bonami, P., Salvagnin, D., Tramontani, A., 2020. Implementing automatic Benders decomposition in a modern MIP solver, in: International Conference on Integer Programming and Combinatorial Optimization. Springer, pp. 78–90.

Bui, M., Adjiman, C.S., Bardow, A., Anthony, E.J., Boston, A., Brown, S., Fennell, P.S., Fuss, S., Galindo, A., Hackett, L.A., Hallett, J.P., Herzog, H.J., Jackson, G., Kemper, J., Krevor, S., Maitland, G.C., Matuszewski, M., Metcalfe, I.S., Petit, C., Puxty, G., Reimer, J., Reiner, D.M., Rubin, E.S., Scott, S.A., Shah, N., Smit, B., Trusler, J.P.M., Webley, P., Wilcox, J., Mac Dowell, N., 2018. Carbon capture and storage: the way forward. Energy Environ. Sci. 11, 1062–1176.

Cho, S., Li, C., Grossmann, I.E., 2022. Recent advances and challenges in optimization models for expansion planning of power systems and reliability optimization. Comput. Chem. Eng. 165, 107924.

Conejo, A.J., Baringo, L., Kazempour, S.J., Siddiqui, A.S., 2016. Investment in electricity generation and transmission, Cham Zug, Switzerland: Springer International Publishing. Springer.

Demirhan, C.D., Tso, W.W., Powell, J.B., Heuberger, C.F., Pistikopoulos, E.N., 2020. A Multiscale Energy Systems Engineering Approach for Renewable Power Generation and Storage Optimization. Ind. Eng. Chem. Res. 59, 7706–7721.

Draxl, C., Clifton, A., Hodge, B.-M., McCaa, J., 2015. The Wind Integration National Dataset (WIND) Toolkit. Appl. Energy 151, 355–366.

eia, 2019. New methanol plants expected to increase industrial natural gas use through 2020.
eia, 2019. Annual Energy Outlook 2019: Energy Prices by Sector and Source
ERCOT, 2016. Hourly Load Data Archives
European Commission, 2019. Communication from the Commission to the European Parliament, the European Council, the Council, the European Economic and Social Committee and the Committee of the Regions :The European Green Deal. European Commission, Brussels.
Fasihi, M., Efimova, O., Breyer, C., 2019. Techno-economic assessment of CO_2 direct air capture plants. J. Cleaner Prod. 224, 957–980.
Flint, D.J., 2004. Strategic marketing in global supply chains: Four challenges. Ind. Mark. Manag. 33, 45–50.
Geoffrion, A.M., 1972. Generalized Benders decomposition. J. Optim. Theory Appl. 10, 237–260.
González-Garay, A., Frei, M.S., Al-Qahtani, A., Mondelli, C., Guillén-Gosálbez, G., Pérez-Ramírez, J., 2019. Plant-to-planet analysis of CO_2-based methanol processes. Energy Environ. Sci. 12, 3425–3436.
IEA, 2020. Projected Costs of Generating Electricity 2020. Paris.
IEA, I.E.A., 2019s. IEA, Electricity generation by fuel and scenario, 2018-2040
Ioannou, I., D'Angelo, S.C., Galán-Martín, Á., Pozo, C., Pérez-Ramírez, J., Guillén-Gosálbez, G., 2021. Process modelling and life cycle assessment coupled with experimental work to shape the future sustainable production of chemicals and fuels. React. Chem. Eng. 6, 1179–1194.
Ioannou, I., Marcandetti, A., Gosálbez, G.G., 2022. Life cycle optimization of energy systems integrated with carbon capture and utilization, 32 European Symposium on Computer Aided Process Engineering. Elsevier, pp. 169–174.
Kostin, A.M., Guillén-Gosálbez, G., Mele, F.D., Bagajewicz, M.J., Jiménez, L., 2011. A novel rolling horizon strategy for the strategic planning of supply chains. Application to the sugar cane industry of Argentina. Comput. Chem. Eng. 35, 2540–2563.
Lara, C.L., Mallapragada, D.S., Papageorgiou, D.J., Venkatesh, A., Grossmann, I.E., 2018. Deterministic electric power infrastructure planning: Mixed-integer programming model and nested decomposition algorithm. Eur. J. Oper. Res. 271, 1037–1054.
Li, C., Conejo, A.J., Liu, P., Omell, B.P., Siirola, J.D., Grossmann, I.E., 2022a. Mixed-integer linear programming models and algorithms for generation and transmission expansion planning of power systems. Eur. J. Oper. Res. 297, 1071–1082.
Li, C., Conejo, A.J., Siirola, J.D., Grossmann, I.E., 2022b. On representative day selection for capacity expansion planning of power systems under extreme operating conditions. Int. J. Electr. Power Energy Syst. 137, 107697.
Liu, S., Pinto, J.M., Papageorgiou, L.G., 2010. MILP-based approaches for medium-term planning of single-stage continuous multiproduct plants with parallel units. Comput. Manag. Sci. 7, 407–435.
NREL, 2021. Electricity annual technology baseline (ATB) data
Sengupta, M., Xie, Y., Lopez, A., Habte, A., Maclaurin, G., Shelby, J., 2018. The National Solar Radiation Data Base (NSRDB). Renew. Sustain. Energy Rev. 89, 51–60.
Towler, G.P., Sinnott, R.K., 2012. Chemical engineering design : principles, practice, and economics of plant and process design. Butterworth-Heinemann, Oxford.
United Nations, 2016. Framework Convention on Climate Change, Paris Agreement.
Zamarripa, M., Marchetti, P.A., Grossmann, I.E., Singh, T., Lotero, I., Gopalakrishnan, A., Besancon, B., André, J., 2016. Rolling Horizon Approach for Production–Distribution Coordination of Industrial Gases Supply Chains. Ind. Eng. Chem. Res. 55, 2646–2660.

661f A Dynamic Optimization Model for Minimizing the Cost of Low-Carbon Industrial Clusters

Mohamed Lameh, Dhabia Al-Mohannadi, Patrick Linke, Yasir Ibrahim

Minimizing the cost of climate action is essential to ensure the economic feasibility of sustainable development. The literature in the field of optimization and process integration provides a variety of methods that allow the cost minimization of the different available CO_2 reduction options. Such methods generally focus on one of the major pathways, such as energy systems (ES) or CO_2 capture, utilization, and storage (CCUS), and provide optimal systems consisting of a restricted set of considered technologies. Providing an integrated strategy for CO_2 reduction requires a more comprehensive consideration of the different available options which allows the investigation of the synergies existing between the major pathways. This work provides a mixed integer linear programming (MILP) model which considers the dynamics of energy sources, products demand, and prices to optimize the sizing and operation of the different plants in an industrial cluster for a given level of emissions allowance. The model considers the integration of power, CO_2, and hydrogen, as well as energy storage, hydrogen storage, and CCUS. This integrated approach allows the consideration of the effect of the dynamics on the levelized costs of electricity, hydrogen, and CO_2 capture, which have major impacts on the profit margins of the considered products. A case study is presented to demonstrate the proposed method is screening for cost-optimal energy mix, hydrogen production pathways, hydrogen utilization opportunities, and CO_2 integration network, given a set level of allowed emissions. The case study considers renewable and fossil-based energy sources, different hydrogen production pathways (grey, green, and blue), and different production plants. The hydrogen and CO_2 integration networks account for the distances between the sources and sinks to determine the optimal allocations.

544f Optimal Deployment Under Uncertainty of Negative Emissions Technologies in the European Union Power System

Valentina Negri, Ignacio Grossmann, Gonzalo Guillén-Gosálbez

Most scenarios that seek to limit global warming below 2°C rely on negative emissions technologies and practices (NETPs) to remove historical and future anthropogenic emissions. However, the cost, potential, and implications of these technologies at scale are still highly uncertain, especially when predictions are made over a long time horizon. In particular, omitting uncertainties in the modeling of NETPs when assessing the technology readiness level and potential carbon dioxide removal (CDR) may lead to suboptimal deployment paths or even infeasible solutions, particularly in long-term planning.

Uncertainty is broadly classified into exogenous, i.e., decision-independent, and endogenous (type 1 or type 2), i.e., when the realization of the uncertain parameters depends on the decision [1]. In process systems engineering, products demand and prices are examples of exogenous parameters commonly modeled as uncertain. In contrast, technical parameters such as process yield are endogenous – type 2 since their realization is contingent on whether a process is performed or not. Different types of mathematical frameworks can be used to account for uncertainties in optimization problems, such as stochastic programming, chance-constrained programming and robust optimization [2].

In the context of NETPs, Grant and co-authors [3] carried out a pioneering study exploring the feasibility of CDR technologies using stochastic programming. They minimize the expected cost of a selected portfolio of technologies that meets the energy demand in the integrated assessment model TIAM-Grantham [3]. Given an expert-informed probability distribution, they model the uncertain CDR potentials of three selected NETPs, finding that emissions reductions should be performed as fast as possible with renewables that are certain today, instead of relying on uncertain future CDR options.

More recently, Galán-Martín et al. [4] developed a fully deterministic model that identifies the optimal deployment pathway to maximize net carbon dioxide removal [4]. The multi-period linear programming RemovAl oPtImization moDel (RAPID) integrates direct air capture (DAC) and bioenergy with carbon capture and storage (BECCS) in the European power mix. BECCS and DAC can be deployed starting from a given year from 2020 until 2100. Additionally, the authors presented confidence intervals by performing an a posteriori uncertainty analysis on economic and emission parameters.

Here, building on the previously developed RAPID, we apply multi-stage stochastic programming accounting for exogenous uncertainty in the electricity demand, which has to be met as an equality constraint, by dropping the assumption of perfect foresight for the modeling timeframe. The model minimizes the power system's total cost, while meeting a net-zero emissions target in 2050 following a carbon neutrality policy. We introduce fixed-charge costs in the objective function by adding binary variables for the capacity expansion of each technology modeled, which gives rise to the mixed-integer linear program RAPIDU (RemovAl oPtImization moDel under Uncertainty). As a first step to assess the feasibility of NETPs in future scenarios, we apply the methodology developed by Apap and Grossmann [1], adapted to our planning problem. Notably, we define non-anticipativity constraints for the first-stage and recourse variables in all the time periods, excluding the last one. Our study considers two possible realizations of the demand – high and low – for each European Union country, resulting in 64 scenarios.

We relax the constraint on the deployment of the NETPs from a given year, letting the model decide when to install the technologies according to the emissions targets. The minimum cost solution features only BECCS as NETPs starting in 2020, while DAC is never deployed in the given time horizon. Capacity expansions take place in subsequent time periods until 2050. The total cost of the power system increases by roughly 10% compared to the case where uncertainties are omitted.

The problem is implemented in the General Algebraic Modeling System (GAMS) software [5] version 35.2.0 and solved with the CPLEX 20.1.0.1 solver on an Intel i9-9900 CPU, 3.10 GHz computer with 32 GB RAM. RAPIDU features 7,871,425 continuous variables, 300,672 binaries and 4,342,346 equations. The solution time is 24,184 s with a 3% optimality gap using parallel threads. The computational time can be further reduced by applying a decomposition method.

This work highlights the advantages of accounting for uncertainties in NETs deployment planning to meet the climate targets. Our analysis can easily be extended to other uncertain parameters, i.e., carbon dioxide storage and biomass availability. To fully characterize the potential of NETPs under uncertainty, an endogenous analysis must be performed as the next step, where the resolution of the uncertainty depends on whether the technology is installed.

References

[1] Apap, R. M., Grossmann, I. E. Models and computational strategies for multistage stochastic programming under endogenous and exogenous uncertainties. Computers and Chemical Engineering 103, 233–274 (2017). http://dx.doi.org/10.1016/j.compchemeng.2016.11.011

[2] Can, L., Grossmann, I. E. A Review of Stochastic Programming Methods for Optimization of Process Systems Under Uncertainty. Frontiers Chemical Engineering 2, 622241 (2021). https://doi.org/10.3389/fceng.2020.622241

[3] Grant, N., Hawkes, A., Mittal, S., Gambhir, A. The policy implications of an uncertain carbon dioxide removal potential. Joule 5, 2593-2605 (2021). https://doi.org/10.1016/j.joule.2021.09.004

[4] Galán-Martín, Á., Vázquez, D., Cobo, S., Mac Dowell, N., Caballero, J. A., Guillén-Gosálbez, G. Delaying carbon dioxide removal in the European Union puts climate targets at risk. Nature Communications 12, 6490 (2021). https://doi.org/10.1038/s41467-021-26680-3

[5] Brooke, A., Kendrick, D., Meeraus, A. andamp; Raman, R. GAMS—A User'sManual. (GAMS Development Corporation, 1998)

17a Process Analyses of Selective NO_x Decomposition Employing Chemical Looping Scheme

Pinak Mohapatra, Sonu Kumar, Rushikesh Joshi, Liang-Shih Fan

Rapid technological advances have led to precipitous rise in energy consumption which have contributed to environmental degradation. The pollutant NO_x generated because of exorbitant use of fossil fuels are primary concern for photochemical smog, acid rain, tropospheric ozone, and ozone layer depletion. This pollutant needs to be removed from the stack gases prior to releasing it into the atmosphere. Additionally, the removal or purification process needs to be sustainable and economical to reduce the plant's operating cost. Hence, NO purification especially at lower temperatures is widely investigated across the globe. In detail, NO direct decomposition is favorable below 1000°C; however, the reaction has a high activation energy barrier of 335 kJ mol^{-1}. Although the current state-of-the-art process, selective catalytic reduction (SCR) exhibits good selectivity and conversion towards NOx purification, it faces several drawbacks such as poor thermal stability, catalyst poisoning due to the presence of other acidic gases in the stack gas like CO_2 and suffers from limited operational flexibility. Additionally, the industrial SCR catalysts are made of noble metals like Pt, Pd, and Rh, which have recently become very expensive thus resulting in an overall increase in the process economics. Furthermore, the operation of SCR requires use ammonia which is expensive when compared to other abundant and cheaper reducing gases like natural gas.

Methane contained in natural gas is a well-known greenhouse gas and is about 25 times more potent than CO_2 at trapping heat in the atmosphere. Due to these reasons, economic and sustainable pathways for natural gas utilization alongside addressing the environmental concerns are recently being explored. This work presents a unique scheme for NOx purification alongside natural gas utilization.

The proposed process is based on a chemical looping scheme that involves carrying out a redox reaction in two different reactors assisted by a metal oxide carrier cyclically looping between two reactors. The process scheme for CLND involves the operation of the NOx purification process operated autothermally. In the first reactor i.e., reducer, the carrier containing an oxidized metal oxide carrier is reduced using natural gas primarily containing methane. It is ensured that the natural gas is completely oxidized to CO_2, with 100% purity. After carrier is reduced, the flue-gas stream containing NO is sent to the reactor, where it selectively reacts with NO to decompose into N2. In addition, the process is studied over a varied temperature range of 300-700 °C and is not affected by components like CO_2 and H_2O in the flue-gas stream. Moreover, the methane utilization for regeneration produces a sequestration- ready pure CO_2 stream. The reaction scheme for CLND is as follows:

Reactor 1(Reducer): $4NiO + CH4 \rightarrow 4Ni + CO_2 + 2H_2O$ (1)

Reactor 2(Oxidizer): $4Ni + 4NO \rightarrow 4NiO + 2N_2$ (2)

Overall: $4NO + CH_4 \rightarrow 2N_2 + CO_2 + 2H_2O$

Process simulations involving sensitivity studies varying oxygen carrier flow rates, operation temperature and inert support composition are conducted. Operating zone of the reducer is chosen based on near 100% methane conversion and CO_2 selectivity alongside no carbon output. Moreover, the reducer and oxidizer are operated in adiabatic and isothermal mode respectively showing the autothermal operating condition. Since reducer is adiabatic and reduction reaction is endothermic, a drop in temperature across the reducer's length is observed. Therefore, it is ensured the reducer outlet temperature is beyond the activation temperature of the carrier to ensure maximum reactivity. To improve heat transfer from the oxidizer to the reducer, addition of inert with higher heat capacity is investigated. Based on exergy analysis, the Effective thermal efficiency (ETE) of the CLND process is calculated ~18 percentage points higher than SCR process.

Besides, the experiments were conducted to determine the activation of CH_4. NO_x reactivity was assessed at different temperatures to determine the best operating temperature for simulation studies. The carrier was tested in a TGA across 20 cycles to assess its redox performance. X-ray diffraction was used to identify the phase changes during the redox process. SEM combined with EDS was performed on fresh and post cycles samples to observe changes in surface morphology. BET analysis showed no significant change in surface area for fresh and post 20 redox cycles. This unique scheme comprising of purifying a pollutant and utilizing natural gas while generating heat for electricity is highly promising.

17c Techno-Economic Analysis of a Combined Power Plant CO2 Capture and Direct Air Capture Concept for Flexible Power Plant Operation

Moataz Sheha, Edward Graham, Dharik Mallapragada, Emre Gençer, Phillip Cross, James Custer, Adam Goff, Ian Cormier, Howard Herzog

Carbon capture and storage (CCS) equipped power plants can reduce the cost of decarbonizing power systems. However, deeply decarbonized power systems also will see significant penetration of variable renewable energy resources that will necessitate flexible operation of other balancing resources in the system such as energy storage and low-carbon dispatchable power plants, including those relying on CCS. Here, we investigate a new concept for power plant flue gas capture integrated with a novel lime-based direct air capture technology in a way that allows for flexible operation of the power plant while having the potential for negative emissions under all operational scenarios. A process simulation and an accompanying techno-economic model is developed to evaluate the capital and operating expenditures and to investigate the economic viability of such concept under different electricity price scenarios.

The proposed concept is composed of three main technologies: calcium looping, membrane + cryogenic separation, and lime-based direct air capture. The system is designed to treat flue gas from a conventional natural gas combined cycle (NGCC) power plant. The flue gas from the NGCC power plant is fed to the calcium looping system which is composed of two reactors: a calciner and a carbonator. Calcium carbonate ($CaCO_3$) is fed to the calciner where it breaks down into lime (CaO) and carbon dioxide (CO_2) at a high temperature. Part of the lime goes to a novel direct air capture (DAC) process that operates in a batch mode to capture additional CO_2 from the air in the form of solid $CaCO_3$ that can be sequestrated or recycled to the calciner. The rest of the lime goes to the carbonator where it captures CO_2 from the power plant flue gas in the form of solid $CaCO_3$ that is recycled to the calciner. The CO_2-rich gas from the calciner goes to a separation system based on commercially available membrane and cryogenic purification technologies to recover high purity CO_2. It is important to note that the DAC process will continue working even when the power plant is not operating. Steady state simulation model for the process is developed in Aspen Plus to investigate the implications of the system under various operational scenarios. Sensitivity analysis is carried out on key process variables to understand the trade-offs associated with the proposed system through detailed process modeling. Results on the process simulation are presented at three different power plant loading levels: 100% (full loading), 40% (part loading), and 0% (shut down). The results show that the system can achieve negative emissions at all power plant loading levels. The results show that the system is net power exporter at full and part loading operational scenarios while being net importer when the power plant is turned off. The developed economic model is used to estimate the levelized cost of electricity (LCOE) under different future scenarios of electricity price, carbon price and natural gas prices.

17d Evaluation of Aprotic Heterocyclic Anion Ionic Liquids for Post Combustion Carbon Capture

Tracy Benson, Adhish Chandra Saketh Madugula

Capturing of waste CO_2, particularly from large industrial point sources (electric power generators, refineries, and other chemical facilities), is necessary for either permanent storage or utilization of CO_2. The technology currently proposed for post-combustion CO_2 capture uses an aqueous solution of 30 wt % monoethanolamine (MEA). However, amine-based solvents have high operational expenses as they are plagued with thermal degradation, degeneration in the presence of impurities (O_2, SO_x, and NO_x), high regeneration energy requirements, and loss of solvent during regeneration. These challenges may be overcome using ionic liquids (ILs) that can be tailored for operating under flue gas stream conditions. ILs, however, suffer from high viscosities, leading to diffusion limitations, compared to amine solvents, and some are ineffective in the presence of water.

A new family of ILs containing aprotic heterocyclic anions (AHAs) and phosphonium cations have been discovered that maintain high CO_2 absorption performance in the presence of water. While they do have significantly higher viscosities that aqueous MEA, their viscosity does not increase with CO_2 saturation. This phenomenon is likely due to the absence of the H^+ ion in the anion, preventing the IL from forming hydrogen bonding. Their viscosities, however, are on the order of 80 to 160 cP, compared to 7 cP of aqueous MEA at combustion capture conditions.

This paper uses Aspen Plus process modeling to evaluate triethyloctylphosphonium 2-cyanopyrrolide, [P2228][2-CNPyr] and trihexyltetradecylphosphonium 2-cyanopyrrolide, [P66614][2-CNPyr], for use as carbon capture absorbents either alone or mixed with various amounts of aqueous MEA. The model used the Electrolyte Non-Random Two Liquid – Redlich Kwong (ENRTL – RK) thermodynamic package to account for the IL/MEA/H_2O, CO_2 electrolyte system. The components were modelled using a True Components Approach that takes into consideration the free and dissociation forms of the components. Our model parameters (viscosity profiles, absorption capacity, enthalpy, and entropy) were validated using laboratory and pilot plant data available in the literature. Our results indicated significant decreases in viscosity when either IL was mixed with aqueous MEA, which reduces mass transfer resistances. The CO_2 loading does not appear to have a large effect on the absorption capacity at lower MEA concentrations (< 30 mole %). The CO_2 – IL absorption is mostly due to chemisorption and is slightly exothermic. However, the heat is readily absorbed by the water, and the absorber temperature can be easily maintained at ~50 °C. Furthermore, compositions and process conditions were optimized to capture ~ 90 % of CO_2 in the flue gas stream (based on an MEA carbon capture process previously published by our group).

17b Recovery of Vapors and Organic Compounds during Industrial Coffee Roasting

Saad Barbar Netto, Murilo Innocentini, Cristina Paschoalato, Silvia H. Taleb-Contini, Tapas Das, Pedro Pires

Brazil is the largest coffee producer and exporter in the world, producing approximately 68 million sacks a year. To supply the required consumption, roasting is done industrially on a large scale. The time and degree of roasting are not only related to the temperature profile of the coffee, but also to higher extraction levels. During roasting, several factors are responsible for the formation of aromatic and volatile compounds. More than 1000 components have been identified, of which almost 850 are volatile aromatic compounds such as furans, pyranes, ketones, and phenols, among others. The volatile compounds are not only present in roasted coffee; during roasting several volatile compounds are lost. Within this context, the present work proposed to evaluate a procedure that allows the recovery, in a semi-industrial scale, of the volatile compounds released to the atmosphere through the stacks and to associate the degree of recovery with the coffee roasting levels through condensation. To execute this work, a partnership was made with a small coffee production company where the sizing of the flavor and vapor recovery system was based. Two gas capture systems were developed to be installed in the chimney with exhaustion capacity of approximately 40 m^3/h. A total of eleven roastings were conducted, nine of arabica beans with 10 kg and two conilon beans with 5 kg. The collection time at the roaster ranged from 8 to 15 minutes. The volume of condensate recovered was approximately 9 to 16 mL/kg of green coffee. The results obtained for the total organic carbon concentration showed values between 0.15 and 0.35 g/kg of green coffee. The pH found in all samples presented acidity with a range of 2.4 to 7.0. Samples were selected for the identification of the components where it was possible to verify for the Arabica-type coffee an amount of approximately 30% of acetic acid, 26% of furfuryl alcohol, 20% of hydroxyacetone and 15% of caffeine. The compounds identified for the conilon-type coffee was approximately 71% acetic acid, 14% caffeine, 12% furfuryl alcohol and 2.5% ethanol.

Keywords: roasting; condensation; volatile aromatic compounds; recovery.

Enhanced Elimination of Carbonaceous Gaseous Elements and Coal Dust Particulate Matters from Flue Gas by Using the Fully Submerged Self-Primed Venturi Scrubber

Subhrajit Mukherjee, Abesh Chatterjee, Bhim Charan Meikap

In recent years, the entire globe is concerned about the drastic effects of hazardous environmental air pollution. Due to several uncontrollable sources of outdoor and indoor air hazards, the environmental norms cannot be maintained at their level of the threshold value. Moreover, the effects of environmental air nuisance are imposing a direct threat to the health of several life forms and the balanced ecosystem existing on our mother planet. The overall air pollution mainly includes hazardous gaseous effluents and solid particulate matter (PM10, PM2.5, PM1). The present century is suffering from the continuous exposure of carbonaceous gaseous matter CO2, CO, and its associated dust particulates. Such target pollutants or effluents can be easily found from the outlets of steel industrial belts, coal mines, sponge iron industries, etc. as industrial flue gas effluents. Besides tracking down the sources, causes, and associated problems arising out of the concerned target material, it is also quite essential to control as well as monitor or control the entire effects of air pollution. The present study mainly details the design of a modified self-primed venturi scrubber for almost complete removal of target material like CO2 from flue gas and its associated coal dust matters. Coal dust particulate matter constitutes among the greatest source of dust-laden mass and can accordingly harm the natural environment or lifeforms in various ways. The modification of the conventional venturi scrubber has been done in the fully submerged self-priming mode to operate in a more economical way along with a higher level of removal efficiency. The self-primed venturi scrubber is primarily composed of the conventional section, throat section, and diverging section along with the outer cylinder mounted around the periphery of the venturi section. Experimentation proves that a higher level of mass transfer as well as removal efficiency is achieved at a higher liquid level or height of liquid solution within the cylinder. Moreover, inlet liquid velocity and concentration of gaseous mixture or concentration of dust-laden particulate matter like coal dust are the governing parameters for calculating or tabulating the overall removal efficiency. An additional comparison study was carried out between fully and partially submerged conditions, which resulted in 50% greater removal for the former as compared with the latter. Considering the fine coal dust matters, 99.90% of removal was noted at the highest liquid inlet velocity of 60 m/s, having an inlet dust concentration of 3 g/m^3 and liquid height of 1m in the outer cylinder. In the case of CO2 emission from industrial flue gases, similarly, a self-primed venturi scrubber is used for its removal at greater heights. 97.5% maximum removal is recorded with an inlet gaseous concentration of 1500 ppm, having a liquid inlet velocity of 80 m/s with a liquid height of 1m within the outer cylinder. The results quite clearly depict that the highest removal was achieved at higher liquid height within the fully submerged condition, maximum inlet concentration of inlet effluent from industrial flue gas, and highest inlet liquid throat velocity or flow rate. The experimentation was successfully run with recycled wastewater or household greywater or industrial makeup water as the scrubbing liquid for maintaining environmental sustainability, zero discharge norms, and the economical prospect of the modified scrubbing device. The air-water molecular level of interaction using the physics of phenomena like atomization, direct interception, gravitation, and centrifugal force of action helps in enhanced removal of target material in an energy-efficient way. As an overall study, the modified scrubbing device is capable of eliminating maximum hazardous carbonaceous gaseous matter and solid particulate matter within a short device residence time using low-cost scrubbing as a water medium. This innovative user-friendly scrubbing technique mainly helps in maintaining or controlling the atmospheric pollution level at its best and accordingly monitoring the ecological balance between nature as well as mankind.

446a Visualization of 35+ Years of Data from the Doe Hanford Nuclear Reservation Site and the Associated Applications for Informing Key Tank Waste Treatment and Disposal Decisions

Courtney Bottenus, Heather B. Sabella, Erin L. McCann, Mark B. Triplett

Big data sets are both voluminous and complex and arise from any single, or a confluence of, large spans (e.g., decades) of data, high frequency data collected over brief intervals (e.g., hours, days, weeks), and/or sophisticated suites of measurement devices and instrumentation (e.g., dozens of instruments) operating and collecting information in parallel and/or sequence. Inherently these large data sets are collected and stored in different forms and formats and access to the data may vary across and among organizations based on access to hardware and software programs, all while multiple authoritative data sources could represent overlapping data information. Therefore, a key challenge becomes integrating this disparate data and eliciting both its value and veracity.

As the U.S. Department of Energy (DOE) Hanford Nuclear Reservation Site, in Richland, Washington, and its contractors (e.g., Washington River Protection Solutions (WRPS) LLC and Bechtel National Inc.) move towards the start-up of Direct Feed Low-Activity Waste (DFLAW) mission operations for the retrieval, treatment, and disposition of tank waste, assuring regulatory compliance must also factor into strategic decisions. As such, key personnel and stakeholders across the Hanford complex (contractors, DOE Office of River Protection (ORP), national laboratories, regulatory agencies, etc.) must have access to a consistently derived set of tank waste characterization data. A strategic suite of Tableau tools has been developed and implemented through the collaborations of Pacific Northwest National Laboratory (PNNL), WRPS, and ORP to more efficiently and consistently enable access and visualization to a common derived "standard" data set and related information to appropriately inform key decisions like technological investments, process decisions, and regulatory actions.

This presentation will showcase several Tableau tools, in the suite, which have been designed and implemented to help inform key decisions associated with the treatment and disposal of Hanford tank waste. While this subset of tools serve as direct demonstrations of the capability needs today, they additionally facilitate the illustration of the breadth, flexibility, and success with this methodology including aspects of data provenance, traceability, and review processes.

446b Temperature Comparisons of Cesium Ion Exchange Performance with Crystalline Silicotitanate on Hanford Tank Waste

Amy Westesen, Emily Campbell, Sandra Fiskum, Truc Trang-Le, Ashley Williams, Andrew Carney, Reid Peterson, Matthew Landon, Kristin Colosi

Crystalline silicotitanate (CST) ion exchanger is planned to be used to remove cesium (^{137}Cs) from the aqueous phase of Hanford tank wastes in preparation for vitrification at the Waste Treatment and Immobilization Plant (WTP). The average bulk supernate temperatures of the majority double-shell tanks (DSTs) in the Tank Farms currently average around 16 °C during the spring and winter months. Column and batch contact testing with simulants and actual tank waste were conducted at temperatures between 16 °C and 35 °C to test the impact of operating temperature on ion exchange processes. With a decrease in temperature, CST capacity for Cs increased while the kinetics of the exchange decreased. Based on these results, a determination of temperature impacts on overall mass transfer will allow future modeling of breakthrough performance at a range of process conditions.

446c Nitric-Glycolic Acid Flowsheet Implementation in the Defense Waste Processing Facility to Reduce Generation of Hydrogen

Wesley Woodham, Dan Lambert, Chris J. Martino, Bill Holtzscheiter, Terri Fellinger, Jeremiah Ledbetter, Dylan Baxter

The Defense Waste Processing Facility (DWPF) at the Savannah River Site (SRS) near Aiken, South Carolina, processes and immobilizes the legacy waste from the production of nuclear weapons materials. Although great progress has been made in processing and vitrifying high level waste (HLW) at SRS, approximately 100 million gallons of radioactive waste remains to be processed.

In this paper, we discuss the new flowsheet that has been developed for use in processing high level radioactive waste to virtually eliminate the generation of hydrogen, the facility's primary flammability risk. The replacement of formic acid with glycolic acid along with optimizing HLW processing will allow DWPF to complete the mission of immobilizing the remaining waste sooner and safer.

Formic acid is an excellent reductant for transition metals, especially mercury. However, noble metals, primarily rhodium and ruthenium, catalyze the destruction of formic acid to produce hydrogen and carbon dioxide. Over 20 potential replacements for formic acid were evaluated, and glycolic acid was chosen as the best reductant to replace formic acid. Over 100 tests with simulants and actual waste have demonstrated that the chemical generation of hydrogen is almost eliminated using glycolic acid as the reducing acid.

DWPF plans to begin processing using the nitric-glycolic acid flowsheet in July 2022. This is a simple change as facility modifications are minimal. However, the use of glycolic acid also leads to significant changes in anion destruction and dissolution of transition metals. It will also require the destruction of residual glycolate in the recycle stream using sodium permanganate. The new flowsheet is arguably the most significant processing change since the radioactive startup of DWPF in 1996. Switching to glycolic acid will allow the facility to lower the air purge used to prevent a flammable atmosphere in addition to other processing changes designed to increase throughput. This paper will discuss the implementation of the new flowsheet in DWPF.

446d Applicability of ATR-FTIR to Measure Dense Slurries: A Hanford Case Study

Rupanjali Gurprasad, Steven Crouse, Stefani Kocevska, Ronald Rousseau, Martha Grover

A waste treatment and immobilization plant (WTP) is being constructed by the US Department of Energy (DOE) at their legacy site, Hanford (Washington, USA), to remediate spent nuclear fuel stored in 177 underground tanks (containing approximately 55 million gallons of waste). Following the removal of key radionuclides, approximately 90% of the tank waste is expected to be immobilized as low-activity waste (LAW)[1]. The planned immobilization includes addition of glass forming chemicals (GFCs) to the waste and vitrifying the resulting mixture into borosilicate glass at 1150 °C using joule-heated ceramic melters (JHCM) in the direct feed low activity waste (DFLAW) process[1]. These glasses containing immobilized low activity waste would then be stored in stainless-steel canisters for storage on site[2,3].

To regulate the process and demonstrate compliance with regulatory and contractual requirements, offline process sampling (of around tens of thousands of samples) and laboratory chemical analysis is planned extensively throughout the DFLAW process[5]. These offline techniques not only expose personnel to the waste thereby causing safety hazards, but are also time consuming, and often require sample preparation which incurs additional time lags, thus failing to provide data representative of immediate conditions in the tank. To counter these challenges, employment of *in-situ* techniques such as infrared (IR) and Raman spectroscopy under the real time in-line (RTIM) program by DOE is planned to facilitate quick decision making by obtaining the required concentration data in real time without any sample preparation.

The current work is focused on studying the applicability of attenuated total reflectance Fourier transform infrared (ATR-FTIR) spectroscopy to monitor slurries expected in the melter feed preparation vessel (MFPV), where the waste is mixed with GFCs. Liquid waste simulants and GFC mixtures designed by Savannah River National Laboratory (SRNL) were used in this study. The simulants are extremely basic (pH>13) and are mostly sodium salts such as nitrate, nitrite, sulfate, carbonate, phosphate, oxalate, chromate, and acetate; the first four being the analytes of most significance for waste processing at Hanford. The GFC mixtures contain up to twelve metal oxides and silicates, but majorly comprise of silica (SiO_2), kyanite (Al_2SiO_5), wollastonite ($CaSiO_3$), and boric acid (H_3BO_3)[11]. The mixing of GFCs with the waste simulants enhances the complexity of the resulting slurry, thereby making monitoring more challenging. Hence, this work aims to define the limits of applicability of attenuated total reflectance Fourier transform infrared (ATR-FTIR) spectroscopy in conjunction with a particle vision and measurement (PVM) probe to monitor thick slurries as a function of turbidity.

The study was conducted in two parts. The first part was focused on understanding the dissolution behavior of individual GFC components in a 3M NaOH solution (to mimic the alkaline nature of the waste). The second part involved monitoring slurries formed after GFC addition to the liquid waste simulants. The IR results show that some GFC components like silica, kyanite, boric acid and lithium carbonate exhibit dissolution in an alkaline medium. Additionally, dissolution of silica and kyanite in solution interferes with the characteristic IR peak of sulfate, while dissolution of lithium carbonate interferes with carbonate and nitrate present in the liquid waste stream. Thus, dissolution of GFC components may hinder accurate monitoring of important analytes in solution.

In addition to studying the dissolution behavior of the GFC components, the effect of the presence of solids (GFCs) on the IR spectra of liquid waste simulants was investigated in the second part of the study. IR was able to identify the characteristic spectral features of key analytes (nitrate, nitrite, sulfate and carbonate) even at GFC loadings as high as 600 gm/L in waste simulants. The turbidity of the slurry resulting after the simulants were loaded with 600g/L of GFCs was observed to be around 0.7 T.U., as compared to 0.11 T.U. of only the liquid simulants. Though, the GFC concentration of 600 g/L tested in this work is the minimum loading of GFCs planned at Hanford, our results demonstrate that the IR signal is unaffected by solids even in such thick slurries, indicating that *in-situ* ATR-FTIR probes can be possible alternatives to offline grab sampling. Thus, this study intends to demonstrate that the employment of online techniques paired with chemometric models might be an effective tool to alleviate the safety hazards posed by traditional offline methodologies.

Bibliography:

(1) Kelly, S. E. A Joule-Heated Melter Technology for the Treatment and Immobilization of Low-Activity Waste. *Washington River Protection Solutions Report RPP-48935* **2011**, *767* (Rev. 0).

(2) Pegg, I. L. Turning Nuclear Waste into Glass. *Physics Today* **2015**, *68* (2), 33–39. https://doi.org/10.1063/PT.3.2687.

(3) Goel, A.; McCloy, J. S.; Pokorny, R.; Kruger, A. A. Challenges with Vitrification of Hanford High-Level Waste (HLW) to Borosilicate Glass – An Overview. *Journal of Non-Crystalline Solids: X* **2019**, *4*, 100033. https://doi.org/10.1016/j.nocx.2019.100033.

(4) Muller. Letter Report : Proposed Approach for Development of LAW Glass Formulation Correlation , VSL-04L4460-1 , Rev . 2. *5*.

(5) Diprete, C. C.; Howe, A. M. WTP Real-Time , In-Line Monitoring Program Task 2 : Determine the Technical Basis for Process Control and Task 5 : Process Control Challenges. **2017**, No. December.

(6) Tse, P.; Bryan, S. A.; Bessen, N. P.; Lines, A. M.; Shafer, J. C. Review of On-Line and near Real-Time Spectroscopic Monitoring of Processes Relevant to Nuclear Material Management. *Analytica Chimica Acta* **2020**, *1107*, 1–13. https://doi.org/10.1016/J.ACA.2020.02.008.

(7) Felmy, H. M.; Lackey, H. E.; Medina, A. S.; Minette, M. J.; Bryan, S. A.; Lines, A. M. Leveraging Multiple Raman Excitation Wavelength Systems for Process Monitoring of Nuclear Waste Streams. *ACS ESandamp;T Water* **2022**, *19*, 42. https://doi.org/10.1021/acsestwater.1c00408.

(8) Kocevska, S.; Maggioni, G. M.; Rousseau, R. W.; Grover, M. A. Spectroscopic Quantification of Target Species in a Complex Mixture Using Blind Source Separation and Partial Least-Squares Regression: A Case Study on Hanford Waste. *Industrial and Engineering Chemistry Research* **2021**, *60* (27), 9885–9896. https://doi.org/10.1021/acs.iecr.1c01387.

(9) Maggioni, G. M.; Kocevska, S.; Grover, M. A.; Ronald, W. Effects of Ion Interactions on the Raman – towards Applications in Spectrum of NO 3 Nuclear Low-Activity Waste at Hanford. **2020**, No. 3.

(10) Dc, V.; Ds, S.; Matyas, K. J. Glass Property Models and Constraints for Estimating the Glass to Be Produced at Hanford by Implementing Current Advanced Glass Formulation Efforts. **2013**.

(11) Siegfried, M. J.; Stone, M. E. LAW Simulant Recipes for Evaluation of Instruments. **2020**.

446f Remediating Contaminated Land on Nuclear Sites

Prashant Mishra, Timothy N. Hunter, Jae Lee, David Harbottle

In the UK, the Sellafield nuclear site has entered a phase of waste management and decommissioning. As part of the decommissioning plan, the site operator will begin to remediate contaminated land on and around the facility to avoid its unnecessary long-term storage as waste. The current research explores the design and use of surfactants to increase the decontamination factor of expandable and non-expandable clays. The study has evaluated the use of cationic surfactants (alkyl ammonium ions) ranging from carbon chain length of 8 to 22 with head groups of methyl, ethyl and propyl for the desorption of Cs^+ from different clay minerals. The decontamination study revealed very good removal of Cs^+ (~98%) from non-expandable clay (kaolinite) using C-22 surfactant. However, decontamination decreased to 70-75% for the expandable clay minerals due to interlayer collapse with Cs^+ adsorbed. Through appropriate design of the surfactant molecule, the intercalation of surfactant led to an expansion of the clay interlayer which improved Cs^+ mobility and increased desorption to ~83.8% and 79.5% for montmorillonite and vermiculite, respectively. The two-phase kinetic model revealed that the desorption of Cs^+ from expandable clay minerals was fast for phase 1 (A1 ~40%) within t1 47 min, which accounted for the outer-sphere complexes, while slow for phase 2 (A2 ~13%) with t2 219 min, attributed to the tightly bound exchangeable ions. However, approximately ~18% of adsorbed Cs^+ was found to be non-exchangeable for desorption due to immobility and fixation in interlayers/FES/hexagon pockets. To enhance desorption performance of the synthesized surfactant, hydrogen peroxide (H_2O_2) was used as an enhancing agent (H_2O_2 catalyzed surface Fe present in clays) and increased the ion-exchange due to its decomposition in interlayers, increasing Cs^+ desorption from expandable clay minerals to ~90%.

193a Novel Sorbents Development for Selective Removal of Pfas from Gaseous Phase and Aqueous Stream

Maoqi Feng

This paper will present two novel sorbents we developed recently for selective removal of per and polyfluoroalkyl substances (PFAS) from gaseous and aqueous phase. Volatile and semi-volatile PFAS contaminants, such as fluorotelomer alcohols (FTOHs), and gaseous byproducts from PFAS treatment, are very challenging. PFAS in groundwater and industrial wastewater are big hurdles for current technologies. The cost-effective sorbents developed overcome the PFAS selectivity challenges, easy to install, and thus provide a unique approach for PFAS removal from both gaseous and aqueous phases. The adsorption capacity of the sorbents was measured with model compounds. The gas phase sorbent can be used to capture PFAS emitted from landfills, and the sorbent for aqueous phase PFAS removal will be combined with the nanofiber membrane technology for landfill leachate treatment. The project was supported by EPA SBIR Phase 2 project (Contract Number 68HERC20C0057).

193b A Silica-Based Approach to Remove Valuable Components from Produced Water

Songpei Xie, Michael Miranda, Anirban Ghosh, Madelyn Shaw, Mark Krzmarzick, Dave Lampert, Clint Aichele

Produced water is a waste stream that is generated during crude oil production. In some cases, produced water contains valuable components such as lithium and lanthanum. Produced water also contains other contaminants such as organics and salt that make recovery of the valuable components challenging. Our work focuses on recovering these valuable components from produced water using a silica-based approach. Our approach leverages the resilient properties of silica such as its high thermal stability and material strength. We evaluated several types of silica, including spherical, fumed, and ligand-modified, to adsorb lithium and lanthanum ions. We also evaluated the adsorption ability of silica in real produced water.

193c Improving Electrode Performance By Engineering *Shewanella Oneidensis* MR1

Jiacheng Zhou, Gregg P. Kotchey, David V. P. Sanchez, Seok Hoon Hong

Microbial fuel cell (MFC) is a device that can treat wastewater and generate electricity simultaneously, which has attracted much attention in recent years. However, several disadvantages, such as the cost of electrode materials, the low buffering capacity of domestic wastewater, and low electron transfer efficiency or low energy production, limit the MFC full-scale application in the industry. One of the main reasons behind the low power density produced by MFC is the slow transfer of electrons from the microorganism to the anode surface. The formation and characteristics of biofilms on the anode of MFC determine the performance of the system. The biofilm that grows on the electrode of the MFC is called electrochemically active, which is the hallmark of the electronic transfer ability with the electrodes. Extracellular electron transfer (EET) is a crucial process by which electricity-producing bacteria transfer biologically generated electrons to extracellular electrodes. *Shewanella oneidensis* has been studied and applied in the biodegradation process of wastewater treatment because of its excellent EET properties. We improved the biofilm formation of *S. oneidensis* by enhancing curli biogenesis and c-di-GMP production. Current density and cyclic voltammetry are measured through a single chamber cell by developing biofilms of our engineered *S. oneidensis* on the commercial glassy carbon electrode. We test efficiency of the bisphenol A degradation and electricity generation with improved biofilm formation of *S. oneidensis* in a single chamber cell. This study will reveal the relation between biofilm-electrode performance and enhanced electron transfer, promoting the application of MFC.

193d Facile Hydrothermal Synthesis of Ferrites and Ferrite/Graphene Nanocomposites for the Catalytic Degradation of Organic Pollutants in Water

Marjorie Lara Baynosa, Amr Hussein Mady, Jae-Jin Shim

Ferrites and ferrite/graphene nanocomposites have been attracting considerable attention in the area of environmental remediation. This study highlights the synthesis and characterization of ferrites and and their composites with reduced graphene oxide (rGO) and their application in the photocatalytic and catalytic degradation of organic compounds in water, specifically methylene blue (MB) and rhodamine B (RhB) dyes and tetracycline hydrochloride (TCH). The magnetically separable ferrite nanoparticles (NPs) and their composites with rGO were prepared in Teflon^{TM}-lined autoclave reactors using a simple hydrothermal method. The samples were characterized using FE-SEM, BET, FTIR, XRD, and XPS. Photocatalytic and catalytic degradation studies of the dyes and TCH were conducted using the synthesized samples. The $ZnFe_2O_4$/rGO nanocomposite exhibited good photocatalytic activity and stability for the degradation of methylene blue under simulated solar light radiation. This result can be attributed to the synergistic interactions between the $ZnFe_2O_4$ NPs and graphene sheets. The $CoFe_2O_4$ NPs and $CoFe_2O_4$/rGO nanocomposite showed excellent removal of RhB and TCH via peroxymonosulfate (PMS) activation to produce sulfate radicals. The synthesized $Zn_xCo_{1-x}Fe_2O_4$ nanoparticles proved more efficient in activating PMS and degrading RhB dye than the pure $ZnFe_2O_4$ NPs, pure $CoFe_2O_4$ NPs, or their physical mixture. Possible mechanisms for the degradation of the organic compounds using the ferrites and composites were proposed. The facile hydrothermal synthesis process used in this study could be used in the synthesis of other new materials. In addition, the synthesis method could possibly be applied to large-scale processes because of its simple-to-follow and environment-friendly nature.

193f Adsorption of Pfas By Zeolites

Charles Ponge, David R. Corbin, Mark B. Shiflett

PFAS are fluorochemicals that have been produced since before the 1940's and have a wide range of uses in consumer and industrial products. While useful, these chemicals have been identified as "forever chemicals", and are ubiquitous almost everywhere. The persistence of PFAS in the environment has called for the monitoring and remediation of drinking water in the US as well as the entire world. While traditional sorbent materials such as activated carbon and ion exchange resins are already being used, they lack the specificity that is needed to work efficiently in large bodies of real polluted water. Zeolites have been identified as potential sorbent materials to selectively remove and concentrate PFAS in the environment. Zeolites have a wide range of pore openings, as well tunable properties to make them ideal for removing pollutants from the environment. Currently there is very little research investigating zeolites for PFAS sorption. In this research, a number of high silica zeolites were investigated for their ability to adsorb PFAS out of water samples. Due to the widespread contamination of low level PFAS everywhere, specific analytical techniques and material considerations had to be used to get consistent and accurate measurements at the ppt level.

662a Microbial Upcycling of Thermally Oxo-Degraded Plastic Wastes

Efrain Rodriguez-Ocasio, Jessica Brown, Robert Brown, Mark Blenner, Laura R. Jarboe

Engineered microorganisms can be instrumental in the fight against plastic wastes and for the advancement of a circular economy. Over 8 billion tons of plastics have been produced, and over half of it has ended or will end in landfills or the environment. Plastic accumulates in the environment because evolution has not had enough time to develop a mechanism to break down plastics into molecules that can be recycled by nature. Instead, it can take a thousand years for plastics to break down via abiotic action, making it the rate-limiting step in plastic biodegradation. Thermal oxo-degradation (TOD) can accelerate this rate-limiting step by rapidly deconstructing the polymer backbone and inserting oxygen functionalities to produce lipids and hydrocarbons. At this point, microorganisms can be employed to convert the carbon that originated from plastic wastes into higher-value products. In this work, non-conventional yeasts were selected against model compounds representative of the composition of thermally oxo-degraded High-Density Polyethylene (HDPE), the most common type of plastic. The down-selected microorganism grew using TOD products from plastics as its sole carbon source without emulsification, overcoming the mass transfer limitations presented by a hydrophobic and solid substrate. The already remarkable growth profile of the microorganism was improved via adaptative laboratory evolution, increasing the growth rate by >200%. This work represents the proof-of-concept for a promising plastic waste upcycling approach and presents a non-conventional yeast with the potential to become a microbial cell factory.

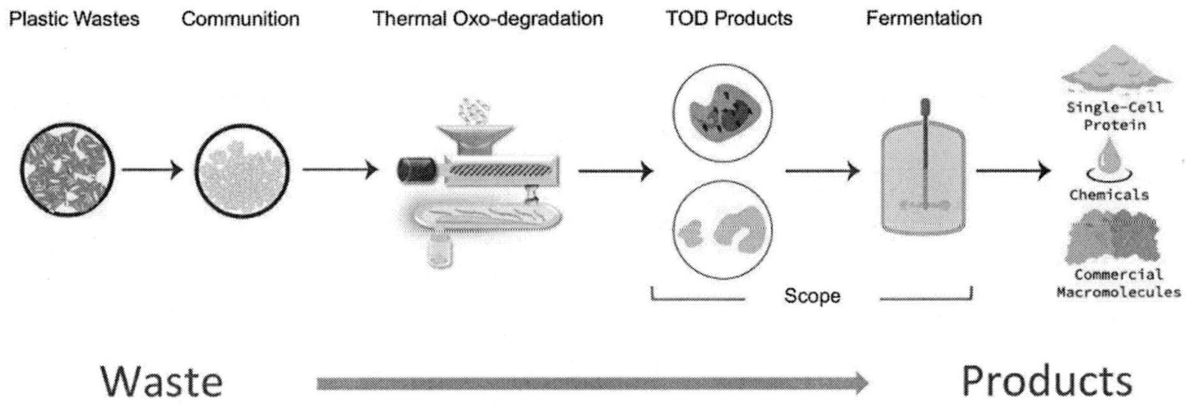

662b Polymer Melt Dynamics on Catalyst Surfaces Using Replica Exchange Molecular Dynamics Simulations

Mehdi Zare, Dionisios Vlachos, Stavros Caratzoulas

Global plastic production has increased exponentially over the last decades and reached over 400 Mt in 2017 (290 Mt of plastic waste every year).[1] Chemical recycling/upcycling has gained significant attention in heterogeneous catalysis to achieve a circular economy. The heterogeneously catalytic reaction of plastics (polymers) occurs at the catalyst-polymer interface. Thus, the structural properties of the polymer on the surface (e.g., which C-C bond is adsorbed) govern the product carbon distribution determining the product. In this context, we developed a methodology using Replica Exchange Molecular Dynamics simulations and the Multi-State Bennett Acceptance Ratio to investigate the structural properties of polymer melts in the adsorbed layer. We applied it to various polyethylene chains and branched polymers. The data indicates that short-chain polymers, e.g., C_{26}, prefer to be fully absorbed, i.e., the energy gained by adsorption overrides the entropic penalty associated with chain propagation along the z direction. Increasing chain length results in a change in the polymer dynamics. Finally, we relate our simulation results with the product distribution we obtain experimentally.

[1] United Nations Environment Programme, *Drowning in Plastics –Marine Litter and Plastic Waste Vital Graphics* (2021).

662c Analysis of Solvent-Assisted Chemical Recycling of Plastics Using a Superstructure Optimization Approach

Austin Lehr, Emmanuel Aboagye, John Chea, Jake Stengel, Kirti Yenkie, Pahola Thathiana Benavides

Introduction

In 2018, the United States generated over 35.7 million tons of plastic waste with only 8.4% recycling rate. The other 91.6% was either incinerated or disposed of in a landfill (USEPA, 2020). By the year 2050 global plastic production is predicted to triple (Tiseo, 2020). Mechanical recycling is the most common recycling method but presents with lower value recycled materials . Chemical recycling presents a promising alternative as it potentially allows for plastics to maintain their original properties (Rahimi and García, 2017). Economic concerns have been the only consideration made regarding the implementation of chemical recycling, however there is the need to incorporate environmental impact assessment for a holistic evaluation. To that end, there is the need to investigate existing and novel feasible chemical recycling technologies to help mitigate the challenging problem posed by plastics end-of-life stage at a cheaper cost, and environmentally friendly. This simultaneous consideration of the technology selection, economics, and environmental burden of plastic recycling presents a multi-objective optimization problem.

Methodology

Thus, we present a conceptual solvent-assisted plastics recycling framework based on a superstructure optimization approach (Chea et al., 2020). Our approach compares different technologies that perform similar tasks based on driving forces, efficiencies, and feed properties. Four crucial stages in this recycling process are implemented into a superstructure: (i) impurity removal, (ii) recovery, (iii) purification, and (iv) refinement. Mathematical models for each technology are developed in terms of mass/energy balances, equipment design and costing and the complete superstructure optimization is formulated as a mixed-integer nonlinear programming (MINLP) problem and solved in General Algebraic Modeling Systems (GAMS) platform. The optimal pathway for the recovery of plastics is then assessed for environmental impacts using the Green-house gas, Regulated Emissions, and Energy use in Technologies (GREET) model (US DOE, 2021.; Wang et al., 2020). In this analysis, the impacts from the upstream processes such as solvent synthesis as well as the recovery process itself are considered. A Life Cycle Inventory (LCI) is developed and calculated for each process and the Life Cycle Assessment (LCA) is a summation of the individual LCIs that make up the process (Laurent et al., 2014).

To test the validity of our framework a case study for the solvent-assisted recycling of poly (ethylene terephthalate) (PET) (Sherwood, 2020) was analyzed. Our framework can simultaneously quantify the cost and the potential emissions of the process with a PET recovery of up to 80% and solvent recovery of 90%. Incorporating solvent recovery into solvent-assisted plastic recycling systems is necessary to make them economically feasible in comparison to traditional mechanical recycling methods.

References

Chea, J.D., Lehr, A.L., Stengel, J.P., Savelski, M.J., Slater, C.S., Yenkie, K.M., 2020. Evaluation of Solvent Recovery Options for Economic Feasibility through a Superstructure-Based Optimization Framework. Ind. Eng. Chem. Res. 59, 5931–5944. https://doi.org/10.1021/acs.iecr.9b06725

Laurent, A., Bakas, I., Clavreul, J., Bernstad, A., Niero, M., Gentil, E., Hauschild, M.Z., Christensen, T.H., 2014. Review of LCA studies of solid waste management systems–Part I: Lessons learned and perspectives. Waste management 34, 573–588.

Rahimi, A., García, J.M., 2017. Chemical recycling of waste plastics for new materials production. Nature Reviews Chemistry 1, 1–11. https://doi.org/10.1038/s41570-017-0046

Sherwood, J., 2020. Closed-Loop Recycling of Polymers Using Solvents : Remaking plastics for a circular economy [WWW Document]. https://doi.org/info:doi/10.1595/205651319X15574756736831

Tiseo, I., 2020. Topic: Global plastic waste [WWW Document]. Statista. URL https://www.statista.com/topics/5401/global-plastic-waste/ (accessed 12.19.20).

US DOE, n.d. Argonne GREET Model [WWW Document]. URL https://greet.es.anl.gov/ (accessed 10.19.21).

USEPA, 2020. Advancing Sustainable Materials Management: 2018 Tables and Figures 84.

Wang, M., Elgowainy, A., Lee, U., Bafana, A., Benavides, P.T., Burnham, A., Cai, H., Dai, Q., Gracida-Alvarez, U.R., Hawkins, T.R., Jaquez, P.V., Kelly, J.C., Kwon, H., Lu, Z., Liu, X., Ou, L., Sun, P., Winjobi, O., Xu, H., Yoo, E., Zaimes, G.G., Zang, G., 2020. Summary of Expansions and Updates in GREET® 2020 (No. ANL/ESD-20/9). Argonne National Lab. (ANL), Argonne, IL (United States). https://doi.org/10.2172/1671788

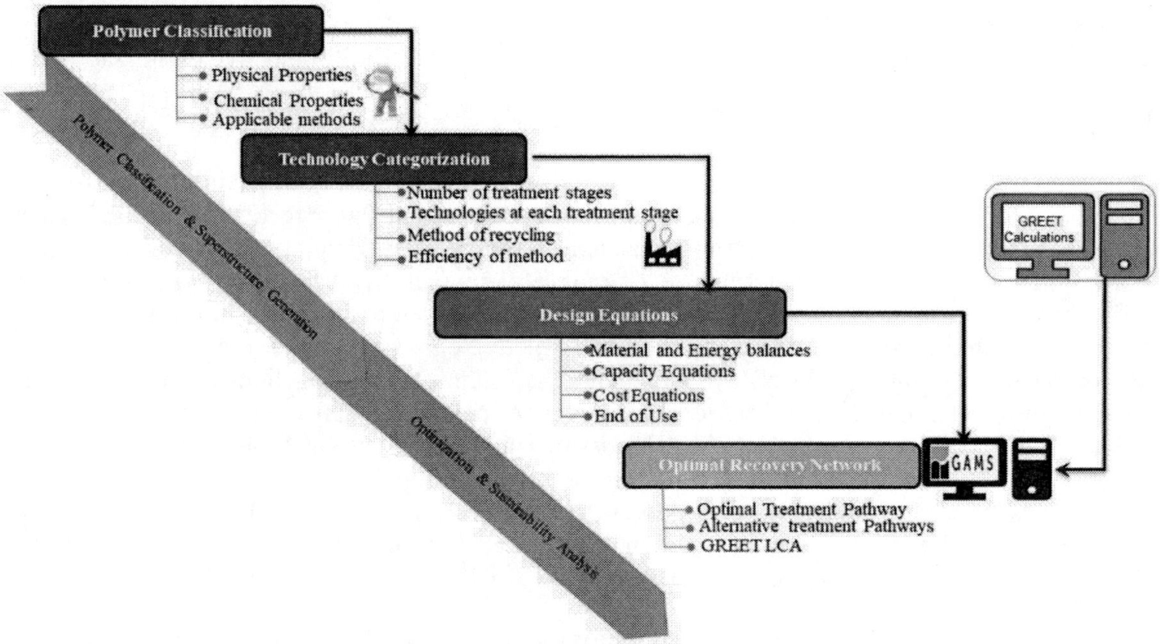

662d Computational Insights into the Thermodynamic Properties of Circular Polymers

Sai Phani Kumar Vangala, Alexander Shaw, Linda Broadbelt

To address end-of-life issues with current synthetic polymers and the lack of proper disposal and recycling strategies, increased attention has been paid to the development of circular polymers that exhibit monomer-polymer-monomer closed loop potential for a circular economy.[1-3] Circular polymers that are produced from renewable resources can be recycled back to their monomer/building block chemicals by overcoming the kinetic barriers of depolymerization.[1-2] This can be achieved by thermal depolymerization at moderate to elevated temperatures (thermolysis) or via the use of catalysts to depolymerize polymers to their monomers in the presence of suitable solvents and moderate temperatures (chemolysis).[2-4] However, the chemical recycling of circular polymers through chemolysis and thermolysis depends on the thermodynamics and kinetics of the de/polymerization. Thermodynamic parameters including the change in enthalpy (ΔH), entropy (ΔS) and free energy (ΔG) govern the thermodynamic equilibrium of de/polymerization. In addition, a concise measure of the de/polymerization, ceiling temperature (T_c), describes the relative rates of the de/polymerization reactions at different working conditions, including monomer concentration.[1-3] The present work focuses on the design of circular polymers through density functional theory (DFT) calculations as a screening method. DFT calculations have been performed to predict the thermodynamic properties (ΔH, ΔS and ΔG) and associated T_c values for oligomers of α-methylene-γ-butyrolacone (MBL), γ-methyl-α-methylene-γ-butyrolacone (MMBL) and α-ethylidene-γ-butyrolacone (EBL). In addition, oligomers of methyl methacrylate (MMA) and MEA were simulated to compare the ΔH, ΔS, ΔG and T_c values with the lactone oligomers.

Gaussian 16 was used to perform geometry optimization and vibrational frequency analyses of the oligomers of MBL, MMBL, EBL, MMA and MEA. The dispersion corrected functional wB97XD along with the 6-31g(d,p) basis set were chosen for the geometry optimization followed by vibrational frequency calculations. Contributions from vibrational frequencies were quantified from -123 °C to 627 °C to evaluate the thermodynamic parameters as a function of temperature and associated T_c. The T_c values for MBL, MMBL, MMA and MEA were found to be 297 °C, 205 °C, 211.5 °C and 112.9 °C, respectively, at 1 M concentration of the monomer. The effects of chain length and the presence of substituents on ΔH, ΔS and T_c values were investigated. Further, a linear correlation has been developed to compare the simulated results with the experimental T_c values, providing an approach to screen various combinations of monomers. This study provides guidelines to experimentalists to synthesize new lactone-based circular polymers that exhibit low to moderate T_c values and an approach that can be extended to other families of polymers to achieve a closed loop circular economy.

References:

[1] C. Shi, L. T. Reilly, S. P.K. Vangala, M. W. Coile, S. R. Nicholson, L. J. Broadbelt, G. T. Beckham, E. Y.-X. Chen, *Chem*, 2021, 7, 1.

[2] R. A. Gilsdorf, M. A. Nicki, E. Y.-X. Chen, *Polym. Chem.* 2020, 11, 4942.

[3] J-B. Zhu, E. M. Watson, J. Tang, E. Y.-X. Chen, *Science*, 2018, 360, 398.

[4] B. A. Abel, R. L. Snyder, G. W. Coates, *Science*, 2021, 373, 783.

662e Additive-Induced Catalyst Poisoning in Polyolefin Hydrocracking

Zachary Hinton, Pavel Kots, Dionisios Vlachos, Thomas H. Epps, III, LaShanda Korley

Functional additives are major impediment to the valorization of plastics waste. For instance, catalytic processes that convert discarded polyolefins into value-added products can enable a range of upcycling opportunities; however, widespread implementation of these technologies is often limited by the fact that plastic waste formulations tend to poison catalysts. Detailed and quantitative analysis of additive effects on catalytic properties (e.g., site chemistry) or process outcomes (e.g., yield, selectivity) is required to guide the design of efficient conversion strategies. In this work, we examine the role of base additive formulation and antioxidant chemistry/concentration on the hydrocracking of high-density polyethylene (HDPE) with a platinum on tungstated zirconium catalyst. We demonstrate a significant reduction (at least 2-fold) in activity between pure polymer and base (additive-containing) resin that is attributable to antioxidants and slip agents. Furthermore, we show that at similar loadings, different antioxidant chemistries have different impacts on catalyst activity and product selectivity. For example, at 0.5 wt% butylated hydroxytoluene (BHT) slows hydrocracking, whereas tri-BHT isocyanurate completely poisons the catalyst. By using infrared spectroscopic studies of antioxidant-treated catalysts, we connect these hydrocracking results to fundamental changes of both functionalities (i.e., metal and acid sites) on the catalyst, which drive varying product selectivities. Overall, this work provides an assessment of one of the key challenges in the deconstruction of 'real' plastics waste and presents an experimental strategy for assessing the performance of potential valorization routes.

662f Chemical Deconstruction of Polyethylene Terephthalate (PET) into Soluble, Nitrogen-Enriched Compounds for Microbial Upgrading

Muhammad Aamir Bashir, Libby Umlor, Laura Schaerer, Stephen Techtmann, Rebecca Ong

At present, plastic is the largest synthetic material produced around the globe; with only <20% plastic recycled, most of it ends up in landfills and open environments. Thermochemical recycling of plastic waste to their monomers has economic limitations as excessive purification and separation processes are required for complete recycling. In this research, we thermochemically generated nitrogen and carbon-rich product from polyethylene terephthalate (PET), avoiding the time and energy costs of direct enzymatic depolymerization prior to microbial upgrading. In the first step, polyethylene terephthalate (PET) was reacted in different concentrations of ammonium hydroxide (4-16%). Reaction conditions were optimized to control the C and N ratio in liquid, including solid loading (0.15-0.3 g/ml), reaction temperature (200-300°C), and time (5-30 min). Under optimized conditions, more than 97% solid PET was successfully converted into the soluble feedstock. FTIR and mass spectrometry techniques were used to characterize the residual solids and liquid products, respectively. The preliminary results suggested; that both natural and engineered microbial consortia can successfully use produced solvents as a sole feedstock to convert plastic into organic biomass as a potential food and feed product.

662g Upcycled Polyvinyl Chloride (PVC) Electrospun Nanofibers Fabricated from Waste PVC for Water Treatment

Atta Ur Razzaq, Milad Esfahani

In recent decades, the increased use of polyvinyl chloride (PVC) in industries and households has led to a surge in PVC waste. PVC is one of the most environmentally harmful plastic that releases toxic chlorine-based chemicals in water, air, and food chain. Upcycling of PVC is difficult owing to a high amount of chlorine and hazardous additives it contains. We report upcycling of PVC-based products as a waste material to a high-value-added product such as membranes by electrospinning technique. In this project, electrospun membranes were fabricated from waste PVC pipe and PVC pool float and their physicochemical properties and performance were compared to membranes produced from commercial PVC powder. The casting solutions for PVC pipe, pool float and commercia PVC powder were prepared by dissolving 15wt.% of each in tetrahydrofuran (THF) and N, N-dimethylformamide (DMF) in a volume ratio of 2:3 respectively. The casting solutions were were electrospun for 6h and membranes were collected on the aluminum foil. The morphology and composition of the membranes was studied using scanning electron microscopy (SEM) which revealed thin and beaded fibers having average diameters of 200.52nm for PVC pipe membranes; in contrast PVC pool float and commercial PVC membranes possessed thicker beadless fibers with average diameters of 281.88nm and 517.42nm respectively. However the membrane generated from the PVC pipe showed thinner thickness (0.038mm) compared to PVC pool float (0.053mm) and commercial PVC (0.065mm) membranes. EDAX studies showed a high carbon and chlorine content in all three membranes. In addition, calcium carbonate () was found in PVC pipe membranes while was negligible in membranes electrospun from PVC pool float and commercial PVC powder. The Young's modulus of electrospun fibers from the upcycled pipe was 73.32 MPa which was significantly higher than that of upcycled pool float (15.63 MPa) and commercial PVC (high modulus: 56.21 Mpa and low modulus:11.83MPa) membranes showing that pipe membrane is rigid while pool float and commercial PVC membranes are relatively elastic. Finally, the adsorption capability of these three membranes was studied for the removal of cadmium (II) (Cd^{2+}) from water and their performance were anlysed based on the physichochemical properties of each membrane.

662h Effective Depolymerization of Polyethylene Plastic Wastes Under Hydrothermal and Solvothermal Liquefaction Conditions

Yixin Liu, Kapil Chandra Akula, Alexa Sanchez, Shuguang Deng

Depolymerization of polyethylene (PE) is one of the most challenging tasks in the chemical upcycling of PE-based plastic wastes because the disassociation of the stable carbon-carbon bonds in PE is only possible at a very high reaction temperature. The thermal liquefaction of PE cable plastic waste in a stainless-steel batch reactor was thoroughly evaluated in this study. The effect of different liquefaction methods (hydrothermal liquefaction (HTL), ionic liquids catalyzed HTL, and solvothermal liquefaction (STL)) on the yields of product fractions (oil products, solid residue, and gas) and the properties of the oil products were examined. At 350 °C and 90 min reaction duration, the conversion rate of 75.43%, the oil yield of 39.33%, the energy recovery rate of 39.7%, the higher heating values (HHV) of 43.83 MJ/kg for the oil samples, and the lower boiling range molecular distribution were obtained by the solvothermal liquefaction method with acetone as a solvent. The HHV of the oil samples obtained in the STL method (43.28-43.83 MJ/kg) is comparable to that of gasoline (HHV - 43.4 MJ/kg). The contribution of the solvent to the depolymerization reaction was mainly the dissolution and dispersion of feedstock by solvation, therefore reducing thermal cracking temperature through enhanced mass and thermal energy transfer. In thermal liquefaction, solvent and feedstock had a low level of solvolysis reactions, so the depolymerization reaction mainly follows thermal cracking. The main reaction path is the random scission of PE molecules during heat treatment, with a low level of polymerization, cyclization, and radical recombination reactions, which occurred through the free radical mechanisms. This work has demonstrated the feasibility of a very promising technique for effective chemical upcycling of polyethylene-based plastics.

692a Multiperiod Design Optimization of a Flexible Power-to-Methanol Process

Andrea Maggi, Jens Bremer, Kai Sundmacher

1. Introduction

Energy transition scenarios require the implementation of process solutions capable of facing the inherent intermittency of renewable utilities. Under the umbrella of Power-to-X technologies, methanol synthesis has received considerable attention in recent years. Methanol is stored in liquid state as a chemical energy carrier or platform chemical, which makes it easily transportable and an ideal candidate for the chemical utilization of renewable energy. While traditional processes for methanol synthesis consider syngas from fossil feedstock (e.g., via gasification of coal, partial oxidation, or steam reforming of natural gas), the CAMERE process feeds CO_2 and H_2, ideally coming from renewable sources. For the CAMERE process, an intermediate reverse water-gas shift (RWGS) reactor partly converts CO_2 into CO before entering the methanol reactor to avoid possible thermodynamic limitations related to the direct hydrogenation of CO_2. Clearly, CO_2 and H_2 constitute important starting materials for methanol in a renewable framework, as the former can be taken from a renewable carbon source (e.g., direct air capture, biogas) and the latter can be produced by water electrolysis, e.g., via high-temperature solid-oxide electrolysis (SOEC). Furthermore, it is relevant to stress the interest in the electrification of endothermic processes to enhance the use of renewable sources. Considering all these aspects, a methanol plant is proposed and optimized with respect to the electrical power supplied by, for instance, an in-house wind farm. The next sections introduce modeling aspects and the optimization setup. Finally, the outcomes of the study are reported.

1. Modeling

The CAMERE process layout extended with a SOEC is reported in Fig. 1. The methanol reactor METHL (I), the core item within the plant, is described as a one-dimensional pseudo-homogeneous bundle reactor, where the coolant temperature is constant along the axis. The reactants are provided via a zero-dimensional SOEC (II) model, where a number of stacks generate a mixed stream of steam and H_2. The carbon feedstock to METHL is supplied by a RWGS isothermal equilibrium step (III) as a mixture of CO and CO_2, while H_2O is condensed and removed (IV) prior to the train of adiabatic compressors with intermediate cooling (V). It is assumed that the catalyst for RWGS is highly selective towards CO, therefore no methane formation is accounted for. A first splitter (VI) directs the outlets from SOEC either to the train of compressors prior to the METHL or to the RWGS. CO_2 can also bypass the RWGS reactor. A second splitter (VII) is assumed to dose the purge-to-recycle ratio after a flash separator (VIII) at the METHL outlets, modeled with the SRK equation of state. The recycle stream is recompressed via a series of adiabatic intercooled steps from the throttling (IX) pressure back to METHL feed. The light green lines starting from the renewable power dashed box (X) represent the electrical power supplied to electrical heaters (in red), SOEC and compressors. The wind farm is solely represented by the available power input W. Cooling heat exchangers are represented in light green.

- optimization setup

All modeling equations were embedded into a NLP optimization problem, where the total power W, supplied by the renewable source (X), is completely utilized within the process. Methanol yield and CO_2 conversion were selected as optimization objectives. Constraints are set to the maximum temperature in the methanol reactor. Important decision variables are METHL and SOEC reactor dimensions, compressor discharge temperatures and pressures, cooling temperature of METHL and operating temperature of RWGS and SOEC. The first optimization problem is based on a single period, where a single period is defined as a fixed power supply Wi, to be completely utilized within the model boundary. The solution of the single period optimization represents a reference solution, which corresponds to the optimal plant setup at a given constant load. Beyond that, the subsequent multiperiod optimization assumed that a number of discrete electrical power loads (periods Wi, i) are representative of the intermittent power supply from renewable sources. The resulting optimal plant design offers a compromise for a set of periods and additionally provides the corresponding optimal operating conditions.

1. Conclusion and future works

Representative scenarios (set of periods) describe the distribution of renewable power loads over time in a highly electrified process for the synthesis of methanol from CO_2 and H_2. The results make it possible to determine the plant design that can best cope with the different loads and to monitor the target deviation over all periods. In future work, the identified design will be a good starting point to establish suitable control strategies for dynamically moving between different load cases. Using the applied methodology will also help to identify flexible process designs for other chemical syntheses to be fed with intermittent renewable electricity.

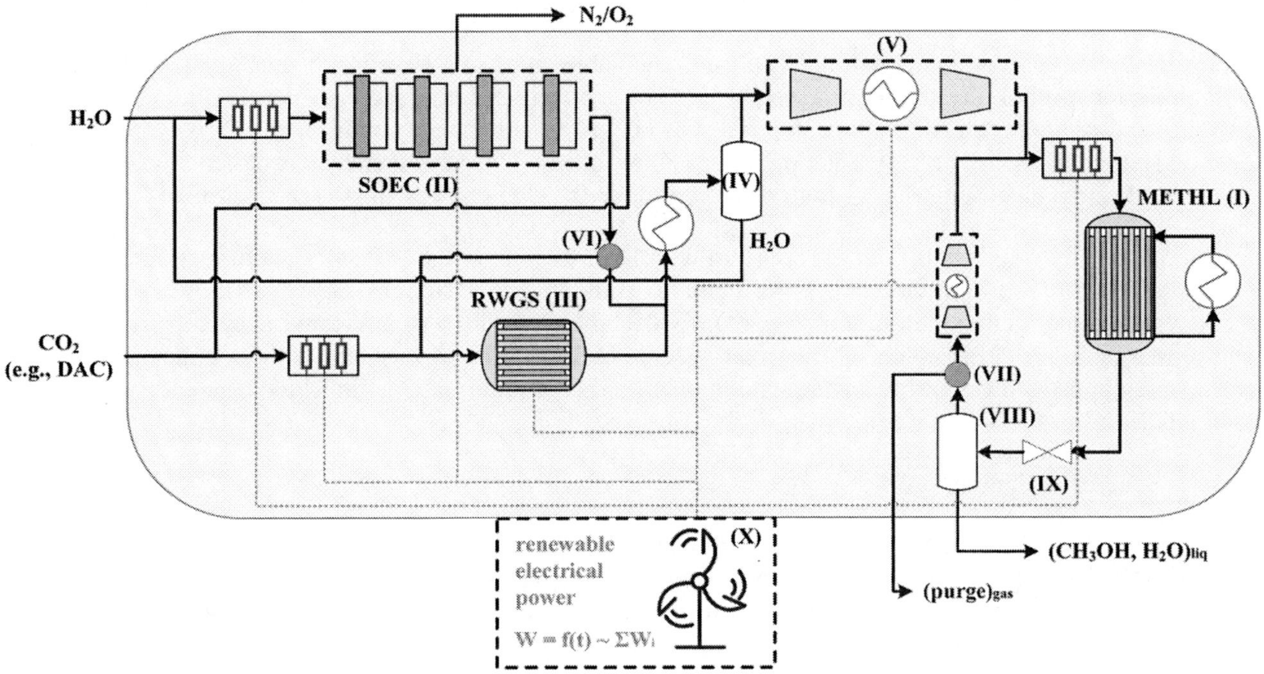

692b Dynamic Modeling and Simulation of a Novel Nuclear-Hydrogen Hybrid Energy System with Large-Scale Storage in an Underground Salt Cavern.

An Ho, Kasra Mohammadi, Matthew Memmott, John Hedengren, Kody Powell

To reduce CO2 and other greenhouse gas emission, electricity grids have been transitioning away from fossil fuel sources to more renewable energy sources. However, technical and economic challenges have risen due to the high intermittency, high variability and low-capacity factor of these renewable energy sources [1], [2]. Power systems with high penetration of renewable energy requires higher flexibility at all times to maintain the balance of supply and demand. Traditional load following plants such as natural gas power plants have been used to meet the fluctuations in electricity demand. These power plants emit significant amount of greenhouse gases [3]. To truly work toward a carbon-free future, these traditional plants have to be replaced by other reliable sources of energy such as nuclear power plants [4].

Integrating nuclear power plants with other clean, renewable energy sources has been shown to increase the flexibility and reliability of the grid system [5]. These hybrid systems increase the flexibility of the grid by storing excess energy during low peak periods as heat (thermal energy storage), electricity (batteries), elevation (pumped hydro), or by producing other commodities. This stored energy can then be used to generate additional electricity during high peak periods with relatively low losses from storage.

A novel nuclear-hydrogen hybrid energy system with large-scale storage in a salt-cavern is proposed as a flexible system for load following [6]. The schematic of the hybrid system is shown in Figure 1. The hybrid system integrates a small modular reactor (SMR) module with a steam Rankine cycle, a polymer electrolyte membrane (PEM) electrolyzer, a large-scale hydrogen storage system with an underground salt cavern, and a hydrogen gas turbine cycle. A dynamic mathematical model is developed for each subsystem in Matlab/Simulink.

Figure 1. Schematic diagram of the proposed NHES

To evaluate the technical feasibility as well as the performance of the hybrid system, a one-year simulation is run using scaled real demand data from ISO New England. The demand data is scaled from a regional profile to a demand profile of a city to account for the size of the nuclear reactor (50 MWe). The results show that the hybrid system's benefit includes reduced nuclear power plant ramping, increased demand met, increased flexibility, as well as reduced greenhouse gas emission.

Due to the addition of the hydrogen generation plant and the storage element in the salt cavern, the hybrid system is shown to be able to load follow while minimizing the load cycles of the nuclear reactor. The flexibility of the system increases due to the addition of these buffers, and enables the nuclear plant to run at base load 92.6% of the year. The number of ramping cycles is also reduced from 533 to 39, a reduction of 92.7%. This helps reduce the maintenance cost as well as the risk of mechanical and thermal fractures on the fuel and cladding of the plant. The addition of the hydrogen generation plant and the gas turbine cycle helps the system meet the peak demands by producing additional electricity during high peak periods. This enables the hybrid system to meet the demand 93% of the year without the need for other electricity sources, compared to just 73.4% of the year in the case of a stand-alone nuclear plant. This means that the grid is less reliant on other fossil-fuel based sources, reducing the amount of greenhouse gas emissions.

Bibliography

[1] M. Sheha and K. Powell, 'Using Real-Time Electricity Prices to Leverage Electrical Energy Storage and Flexible Loads in a Smart Grid Environment Utilizing Machine Learning Techniques,' *Processes,* vol. 7, 2019.

[2] M. Sheha, K. Mohammadi and K. Powell, 'Techno-economic analysis of the impact of dynamic electricity prices on solar penetration in a smart grid environment with distributed energy storage,' *Applied Energy,* vol. 282, 2021.

[3] J. S. Kim, K. M. Powell and T. F. Edgar, 'Nonlinear Model Predictive Control for a Heavy-Duty as Turbine Power Plant,' in *American control conference*, Washington D.C., 2013.

[4] F. J. Sisternes, J. D. Jenkins and A. Botterud, 'The value of energy storage in decarbonizing the electricity sector,' *Applied Energy,* 2016.

[5] J. D. Jenkins, Z. Zhou, R. Ponciroli, R. B. Vilim, F. Ganda, F. de Sisternes and A. Botterud, 'The benefits of nuclear flexibility in power system operations with renewable energy,' *Applied Energy,* 2018.

[6] A. Ho, K. Mohammadi, M. Memmott, J. Hedengren and K. M. Powell, 'Dynamic simulation of a novel nuclear hybrid energy system with large-scale hydrogen storage in an underground salt cavern,' *International Journal of Hydrogen Energy,* vol. 46, no. 61, pp. 31143-31157, 2021.

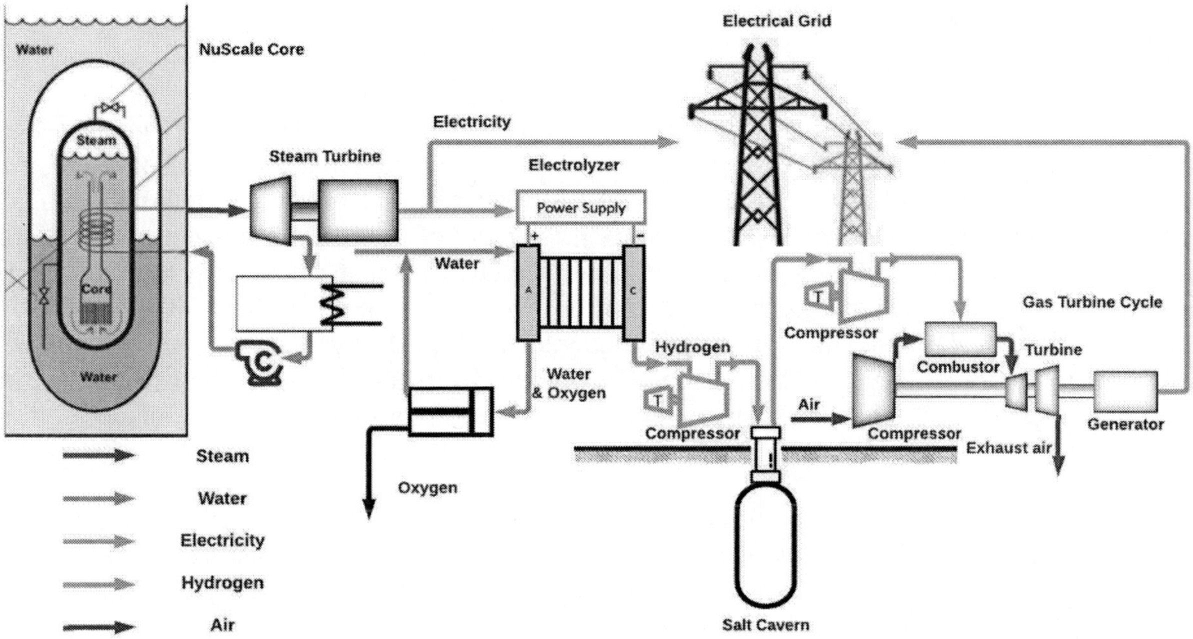

692c Technoeconomic Analysis and Optimization of Low Carbon, Reforming-Based Integrated Energy Systems for the Co-Production of Hydrogen and Power

Maojian Wang, Jinliang Ma, Eric Lewis, Paul Myles, John Brewer, Dale Keairns, Anthony P. Burgard, David Miller

Hydrogen has a strategic importance in the pursuit of a decarbonized, more sustainable energy sector. Integrated Energy Systems (IES) based around hydrogen production and use have the potential to synergistically provide multiple services such as electricity, high value heat, and chemicals as part of dynamic, tightly coupled hybrid systems. This work explores combining a natural gas turbine (GT), a steam methane reforming (SMR) process, and carbon capture to create an IES capable of generating nearly carbon-free hydrogen and power. The GT provides electricity for on-site use, thereby eliminating the need for a stable external power source, and for sale as a co-product when electricity prices are high. In comparison to standalone systems for power and hydrogen generation, the IES concept enables significant savings in capital and maintenance costs due to the sharing of much of the same equipment, such as the carbon capture and compression system, as well as lower variable costs through process and heat integration.

Two IES process configurations – an SMR with a natural gas simple cycle (NGSC) and an SMR with a natural gas combined cycle (NGCC) – are optimized and analyzed using the open-source Institute for the Design of Advanced Energy Systems (IDAES) Integrated Platform [1-3]. In both cases, oxygen-rich flue gas exhaust from the GT is fed to the burners of the SMR furnace in place of air. In the SMR + NGSC design, the flue gas leaving the furnace enters a heat recovery system that preheats the natural gas to be reformed and generates steam for process use. In the SMR + NGCC configuration, the flue gas also passes through a heat recovery steam generator and steam turbine to generate additional power. Flue gas from each system then enters a solvent-based post-combustion carbon capture system as well as a compression and purification unit before being transported through a pipeline for long term geologic storage.

In comparison to standalone systems for power and hydrogen, the IES designs achieve higher thermal efficiencies by recovering the sensible heat contained in the gas turbine exhaust. Additionally, the total amount of flue gas for the IES designs is over 40% lower than standalone systems resulting in smaller carbon capture equipment. Such advantages result in significantly lower levelized cost of hydrogen (LCOH) ($/kg) compared to a standalone SMR with carbon capture [4] after considering a credit for electricity sales. The impact of several economic parameters such as natural gas price, electricity price, carbon price, and capacity factor are analyzed to more fully understand the value proposition of the IES designs. Deployment opportunities are analyzed in the context of a future regional electricity market by coupling the process with locational marginal price signals using a price taker assumption. As expected, the benefit of the IES becomes more pronounced at higher electricity prices, thereby insulating the system from potential future scenarios where the price of electricity is high.

Disclaimer

This report was prepared as an account of work sponsored by an agency of the United States Government. Neither the United States Government nor any agency thereof, nor any of their employees, makes any warranty, express or implied, or assumes any legal liability or responsibility for the accuracy, completeness, or usefulness of any information, apparatus, product, or process disclosed, or represents that its use would not infringe privately owned rights. Reference herein to any specific commercial product, process, or service by trade name, trademark, manufacturer, or otherwise does not necessarily constitute or imply its endorsement, recommendation, or favoring by the United States Government or any agency thereof. The views and opinions of authors expressed herein do not necessarily state or reflect those of the United States Government or any agency thereof.

References

[1] D. C. Miller, J. D. Siirola, D. Agarwal, A. P. Burgard, A. Lee, J. C. Eslick, B. Nicholson, C. Laird, L. T. Biegler, D. Bhattacharya, N. V. Sahinidis, I. E. Grossmann, C. E. Gounaris and D. Gunter, 'Next Generation Multi-Scale Process Systems Engineering Framework,' *Computer Aided Chemical Engineering*, vol. 44, pp. 2209-2214, 2018.

[2] A. Lee, J. H. Ghouse, J. C. Eslick, C. D. Laird, J. D. Siirola, M. A. Zamarripa, D. Gunter, J. H. Shinn, A. Dowling, D. Bhattacharyya, L. T. Biegler, A. P. Burgard and D. C. Miller, 'The IDAES process modeling framewrok and modeling library - Flexibility for process simulation, optimization and control,' *Journal of Advanced Manufacturing and Processing*, vol. e10095, no. https://doi.org/10.1002/amp2.10095, 2021.

[3] https://idaes-pse.readthedocs.io/en/stable/.

[4] E. Lewis, S. McNaul, M. Jamieson, M.S. Henriksen, H.S. Matthews, L. Walsh, J. Grove, T. Schultz, T.J. Skone, and R. Stevens, "Comparison of Commercial, State-of-the-Art, Fossil-Based Hydrogen Production Technologies," DOE/NETL-2022/3241.

692e Conceptual Design and Analysis of a Power Generator with Integrated Thermal Energy Storage and CO_2 Capture

Naresh Susarla, Edna Rawlings, Jaffer Ghouse, John Siirola, David Miller

Conceptual design of an integrated energy system (IES) involving a thermal power generator (TPG), a thermal energy storage (TES) system, and a CO_2 capture system (CCS) is studied for its operation under fluctuating electricity demand. Determining an efficient and economic design for such an IES (i.e., TPG + TES + CCS) is increasingly important for improving operational flexibility, decarbonizing the electricity grid, and reducing the curtailment of intermittent renewable power. However, the design and integration of TES and CCS systems with a TPG is non-trivial and requires the evaluation of several discrete design choices and their effect on the overall power plant operation [1, 2]. Thus, the focus of this study is to use equation-oriented rigorous process models and advanced design optimization methods to determine the optimal IES design using historical electricity price signals with a price taker assumption. The modeling and optimization are done using the Institute for the Design of Advanced Energy System's (IDAES) open-source computational platform [3, 4].

In this work, a 2-step approach is proposed for the IES design. In step 1, a superstructure model is presented, which includes both discrete and continuous design decisions. The discrete decisions include selection of a storage medium for the TES system, steam/condensate sources for operating charge and discharge cycles of the TES system, and a steam source for operating the CCS reboiler. The continuous decisions include heat exchanger areas, fluid flowrates in the TES system, and steam flow rates to CCS. This superstructure model for conceptual design is implemented as a generalized disjunctive programming (GDP) problem with an objective of minimizing the total annualized cost and is solved using GDPopt [5] in IDAES. In step 2, the performance of the optimal IES design is evaluated by solving an operational scheduling problem using a nonlinear model, which is formulated by fixing the optimal design decisions from step 1 in the design superstructure. Additional operating constraints for power plant ramping, monitoring storage inventory, and updating current state of charge are added to the scheduling model. The optimal operating schedule under a price taker assumption is determined by solving the nonlinear problem using IPOPT in IDAES to maximize the operating profit. In our preliminary analysis, for a TES system with a storage capacity of 900 MWh, the optimal IES design is identified from 40 alternatives in step 1. The results from step 2 provide insights into storage utilization, including salt inventory used and maximum heat duties of storage heat exchangers over the scheduling horizon and their sensitivity to the electricity prices. These insights are used to evaluate the assumptions made in the design stage (i.e., step 1).

Acknowledgements

This research was conducted as part of the Institute for the Design of Advanced Energy Systems (IDAES) and Design and Optimization Infrastructure for Tightly Coupled Hybrid Systems (DISPATCHES) projects. It was supported by (1) the Simulation-Based Engineering, Crosscutting Research Program within the U.S. Department of Energy's Office of Fossil Energy and Carbon Management and (2) the Grid Modernization Initiative of the U.S. Department of Energy as part of its Grid Modernization Laboratory Consortium, a strategic partnership between DOE and the national laboratories to bring together leading experts, technologies, and resources to collaborate on the goal of modernizing the nation's grid.

Disclaimer

This project was funded by the U.S. Department of Energy, National Energy Technology Laboratory an agency of the United States Government, through a support contract. Neither the United States Government nor any agency thereof, nor any of its employees, nor the support contractor, nor any of their employees, makes any warranty, expressor implied, or assumes any legal liability or responsibility for the accuracy, completeness, or usefulness of any information, apparatus, product, or process disclosed, or represents that its use would not infringe privately owned rights. Reference herein to any specific commercial product, process, or service by trade name, trademark, manufacturer, or otherwise does not necessarily constitute or imply its endorsement, recommendation, or favoring by the United States Government or any agency thereof. The views and opinions of authors expressed herein do not necessarily state or reflect those of the United States Government or any agency thereof.

References

[1] Susarla, E. S. Rawlings, J. Ghouse, C. L. Laird, M. Zamarripa, M. Bynum, J. Siirola and D. Miller, 'Conceptual design of molten-salt storage system for supercritical power plants,' in 45th Clearwater Clean Energy Conference, Clearwater, Florida, 2021.

[2] S. Rawlings, N. Susarla, J. Ghouse, M. Zamarripa, C. Laird, J. Siirola, M. Bynum and D. Miller, 'Conceptual Design Via Superstructure Optimization in Advanced Energy Systems Using Idaes,' in AIChE Annual Meeting, Boston, MA, 2021.

[3] D. C. Miller, J. D. Siirola, D. Agarwal, A. P. Burgard, A. Lee, J. C. Eslick, B. Nicholson, C. Laird, L. T. Biegler, D. Bhattacharya, N. V. Sahinidis, I. E. Grossmann, C. E. Gounaris and D. Gunter, 'Next Generation Multi-Scale Process Systems Engineering Framework,' Computer Aided Chemical Engineering, vol. 44, pp. 2209-2214, 2018.

[4] Lee, J. H. Ghouse, J. C. Eslick, C. D. Laird, J. D. Siirola, M. A. Zamarripa, D. Gunter, J. H. Shinn, A. Dowling, D. Bhattacharyya, L. T. Biegler, A. P. Burgard and D. C. Miller, 'The IDAES process modeling framewrok and modeling library - Flexibility for process simulation, optimization and control,' J Adv Manuf Process, vol. e10095, no. https://doi.org/10.1002/amp2.10095, 2021.

[5] Q. Chen, E. S. Johnson, J. D. Siirola and I. E. Grossmann, 'Pyomo.GDP: Disjunctive Models in Python,' *Computer Aided Chemical Engineering,* vol. 44, pp. 889-894, 2018

692f Integrated Design and Operation of Low-Carbon District Heating Systems

Gustavo Campos, Nael El-Farra, Ahmet Palazoglu

Electrification has been identified as a major pathway for promoting deep sectoral decarbonization. This is particularly important for sectors historically difficult to decarbonize such as heat generation, and especially relevant under the recent energy insecurity issues related to fuel price and availability caused by political disruptions. For instance, industrial heating is estimated to comprise approximately 20% of global energy demand (IRENA, 2019), while in the US an estimated 2/3 of the industrial thermal demand is required at temperatures lower than 250 °C, ideally suited to the application of renewable heat systems (McMillan *et al.*, 2021). District heating plants centralize heat production and thus present a high potential for increasing energy efficiency and reducing emissions. In this work, we propose a formulation for the integrated design and operation of low-carbon district heating plants including electrical heat pumps, non-concentrating solar thermal collectors, and thermal energy storage tanks, augmented by natural gas boilers. A monolithic multistage modeling approach is adopted, considering both hourly and seasonal variations of ambient conditions and utility prices. The relationship between supply temperature and available heat transfer loads for each equipment is carefully analyzed and equipment models are judiciously selected to retain a linear framework, which is important to ensure that the large-scale integrated design and operation problem remains tractable. Results from a case study identify the potential of each low-carbon technology and highlight their synergistic behavior for different sites in the United States.

References

IRENA. 2019. Innovation landscape brief: Renewable power-to-heat, International Renewable Energy Agency, Abu Dhabi.

McMillan, C., Schoeneberger, C., Zhang, J., Kurup, P., Masanet, E., Margolis, R., Meyers, S., Bannister, M., Rosenlieb, E., Xi, W. 2021. Opportunities for Solar Industrial Process Heat in the United States. NREL/TP-6A20-77760. National Renewable Energy Laboratory, Golden, CO. Available at: https://www.nrel.gov/docs/fy21osti/77760.pdf.

392a Lab- and Pilot-Scale Sulfate-Reducing Bioreactors Treating Acid Mine Drainage from an Abandoned Nevada Gold Mine

Thomas Kaps, Sage Hiibel

Acid mine drainage (AMD) is a global environmental hazard that is produced when water flows over exposed rocks containing sulfur-bearing minerals. AMD has many sources and is produced by active and abandoned mines, leading to a need for effective, low-maintenance and low-cost remediation solutions to prevent the contamination of groundwater, the corrosion of infrastructure, and the disruption of reproduction and growth cycles in plants and animals. Sulfate-reducing bioreactors (SRBR) are currently one of the most cost-effective and low-maintenance solutions for treating AMD. Anaerobic sulfate reducing bacteria take in sulfate as a terminal electron acceptor for their metabolisms to create H_2S, provide energy for the bacteria, and lower the pH of the water resulting in the precipitation of metal sulfides.

In this work, four pilot-scale SRBR were installed as in-field bioreactors in Perry Canyon, NV near the abandoned Jones-Kincaid adit. In parallel, eight lab-scale SRBRs were operated at the University of Nevada, Reno. Each SRBR contained organic substrate (corn stover, pine shavings, and dairy manure), pea gravel to maintain porosity, and a microbial inoculum. The inoculum was obtained from either the anoxic soil of a nearby lake environment (the Sparks Marina) or from the AMD-impacted ephemeral stream of Perry Canyon, with both demonstrating high sulfate-reducing performance in past work. The pilot-scale SRBR were fabricated as 115-L upflow drums and were installed below ground to modulate environmental conditions and were fed AMD directly from the adit. The lab-scale SRBR were 2-L upflow columns fed synthetic AMD mimicking the Perry Canyon AMD composition.

Sulfate and metals concentrations of feed and effluent from each SRBR were monitored temporally. Although results to date indicate limited sulfate reduction occurring in the field-scale SRBR during the first six months of the in-field operation, corresponding to October through April,

the reactors inoculated with the marina soil have slightly greater sulfate reduction than those inoculated with AMD-impacted soil. We believe the colder temperature averages (0 to 10 ^{o}C) in these months are responsible for the lower performance thus far, and we expect performance to increase with increasing spring and summer temperatures to levels comparable to those observed in the lab.

The concurrent operation of field and lab SRBR in this work will provide valuable knowledge of the scale-up process of the treatment technology, as well as insight into how environmental operating conditions may impact the SRBR performance. If designed and operated properly, SRBR have to the potential to be a cost-effective option for AMD remediation at locations around the world.

392b Weathering Induced Changes in Microplastic Dispersity and Pollutant Uptake Capacity

Philip J. Brahana, Ahmed Al Harraq, Bhuvnesh Bharti

Microplastics have become ubiquitous in nature and a part of our everyday life, however their environmental fate remains poorly understood. As microplastics age and endure weathering, their surface chemistry can be altered. An accelerated weathering experiment was performed on microscopic polyethylene particles in the laboratory over a 10-day period. During this period, a change in wettability was observed, which resulted in a shift of the particles to a hydrophilic state and an overall increase of their dispersal in the water column. This change in wettability can be attributed to the destruction of hydrophobic surface groups through the photo-degradation process. To study changes in adsorption, we compared the uptake of two common pollutants onto the microplastic substrate: (1) positively charged malachite green molecules and (2) negatively charged arsenite molecules. We observed an increase in the adsorption of malachite green molecules onto weathered particles, but no significant adsorption of the arsenite molecules throughout the weathering process. This is because during weathering, the magnitude of the negative charge on the PE particles and thus results in electrostatic attraction and repulsion between the malachite green and arsenite molecules, respectively. This finding provides insight on both the transport and dispersibility of MPs in the environment, as well as further understanding their toxicity to human health.

Review of the Fenton oxidation process for industrial wastewater treatment

Amany M. Naguib[1], Soha A. Abdel-Gawad[2], Mohamed K. Mostafa[3], Robert W. Peters[4], Ahmed S. Mahmoud[5]

[1]*Sanitary and Environmental Institute (SEI), Housing and Building National Research Center (HBRC), Egypt*

[2]*Chemistry Department, Faculty of Science, Cairo University, Giza, Egypt*

[3]*Faculty of Engineering and Technology, Badr University in Cairo (BUC), Cairo, Egypt*

[4]*Department of Civil, Construction and Environmental Engineering, University of Alabama at Birmingham, Birmingham, AL, USA*

[5] *Institute of Environmental studies, Arish University, North Sinai, Egypt*

[6]*Scientific Research Development Unit, Egyptian Russian University (ERU), Badr, Egypt*

Abstract

Industrial wastewater is produced in large quantities in developing countries that cannot be treated using typical treatment technologies. Separators and traditional chemical treatment procedures cannot remove partially soluble organic pollutants such as pesticides, PAHs, oils, and grease. High-molecular-weight organic chemicals include dyes and high wastewater output limit chemical treatment techniques. Also, biological treatment methods are limited by the absence of sufficient space Pharmaceutical wastes or harmful organic effluents like phenol cannot be treated by biological methods. The ability of Advanced Oxidation Processes (AOPs), in example Fenton oxidation, to quickly and economically remove a wide range of industrial wastewater contaminants. The Fenton process uses iron to reproduce hydroxyl radicals ($HO^{\cdot}$) produced by hydrogen peroxide. The OH that is created is capable of degrading organic molecules. Fenton's oxidation process may be divided into two categories: conventional Fenton oxidation (CF) and heterogeneous Fenton oxidation. Fenton oxidation can degrade around 87.5±6.67 of industrial COD pollutants in acidic conditions with a pH of 3.6±0.9. Fenton reagents, individually or combined by microwave or UV, can be remove 82 % of pesticides, 98.3% of phenols, 88 % of turbidity, 99 % of Methelene Blue, 100 % of anionic dyes, 87 % of leachate, 68 % of textile effluents, and 95 % of laundrette effluents in short periods of time ranging from 0.25 to 4 hours. Finally, this research found that Fenton oxidation procedures may be employed to remediate industrial wastewater.

1. Introduction

Water is the basis of all life (Wang et al., 2022). The world's population is growing, which results in increased water usage and daily wastewater disposal (Farag et al., 2018, Mahmoud et al., 2021b). Governments are always looking for new methods to save lives and preserve water (Mahmoud and Mahmoud, 2021). The key to solving the water shortage and reduce contaminations is through wastewater treatment and reuse (El-Shafei et al., 2016). There are several types of wastewater that can be introduced into the municipal system, such as household, hospital food industry, metal industry, and painting industry wastewater, and each of them requires a specific treatment process(Mahmoud et al., 2017, Abdellatif et al., 2021).

There are several methods available for treating wastewater, including physical, biological, and chemical approaches (Mahmoud et al., 2018a, Mahmoud et al., 2021a). The most popular physical treatment methods are temperature control, oil separation, and physical separation of large suspended particulates (Mohamed et al., 2016). Also, biological, physicochemical, and chemical treatment can be used for traditional municipal wastewater (Saryel-Deen et al., 2017). Biological treatment, which also necessitates a substantial financial investment, is ineffective for treating industrial effluent (Mostafa et al., 2017). Wastewater from the pharmaceutical, agrochemical, and petroleum refinery sectors is resistant to biodegradation and is difficult to treat effectively using traditional techniques (Mahmoud et al., 2019, Mahmoud et al., 2018b). Chemical treatments take into consideration adsorption techniques, advanced oxidation processes (AOPs), coagulation-flocculation techniques, and Nanotechnology (El-Naas et al., 2010, Mahmoud et al., 2020).

One of the advanced oxidation techniques includes Fenton, and Fenton-based reaction treatment processes are considered to be responsible for the removal of numerous dangerous organic pollutants from the ecosystem (Matavos-Aramyan and Moussavi, 2017). Advanced oxidation processes (AOPs) are classified as chemical, photochemical, catalytic, photocatalytic, mechanical, and electrical reactions because hydroxyl radicals (OH) are created by chemical, electrical, mechanical, or radiation energy. where advanced oxidation methods have become more popular lately. Because all of the solid parts of effluent can be treated by photo-oxidize the wastewater, H_2O_2 or a combination of H_2O_2 and UV is used. There is a largely dependent economy on hydroxyl radicals (OH•) because they are the source of oxidation energy and produce little waste.

The various factors that influence the chemical oxidation of dyes in their aqueous solutions: are pH, hydrogen peroxide (H_2O_2) dose, ferrous sulfate ($FeSO_4.7H_2O$) dose, time, and initial concentration. According to the experiment results, Fenton's oxidation process effectively reached very good elimination effectiveness of over 95%. H_2O_2 and ferrous iron are combined to form Fenton's reagent, which produces hydroxyl radicals. Hydroxyl radicals are produced as a result of the breakdown of H_2O_2, which is initiated and catalyzed by ferrous iron (Fe^{++}) and led to the rise of these radicals. The Fenton reaction, which has a significant oxidation potential, has only lately started to be used in wastewater treatment. Also, UV radiation leads to the production of hydroxyl radicals and the reduction of Fe (III) (Ebrahiem et al., 2017). The major goal of this study was to investigate the feasibility of applying the Fenton oxidation method to reduce the residual organic contaminants that existing in wastewater.

2. Material and Methods

Several Fenton reagents were used by authors during previous investigations and are made by: ferrous chloride extra-pure (hydrate) ($FeCl_2.XH_2O$, 98%), ferrous sulfate heptahydrate catalyst ($FeSO_4.7\ H_2O$), hydrogen peroxide (H_2O_2, 50%), and hydrogen peroxide (H_2O_2, 30% w/w). Also, extra pure materials like silver sulfate, mercuric sulfate and potassium dichromate were used. The change in pH was adjusted using 0.1 M NaOH and 0.1 M HCl solutions in most treatment studies; distilled water was used during previous researches.

3. Result and discussions

The Fenton reaction is applied in acidic conditions to prevent iron precipitation. A complex reaction occurs in an aqueous solution to produce radicals. The efficiency of the treatment was determined by changes in the chemical oxygen demand (COD) of the wastewater. Fenton's reagent experiments were performed under the optimal conditions of Fe^{3+} = 40 mg/L, H_2O_2 = 400 mg/L, and pH=3, As a result of the previous optimum condition 95% of the COD was eliminated within 15 minutes. To prevent the precipitation of iron hydroxide, the pH of the

Fenton reaction must be carried in acidic conditions pH=3 (Tony et al., 2016). The effluents treated by advanced oxidation technology reduced the amount of organic contaminants represented by COD by 70–80% in comparison to biological treatment's COD reduction of 30-45 % (Ghaly et al., 2014).

The Fenton method has an advantage over both coagulation and oxidation mechanisms. While Fenton's reaction makes organic compounds decompose, it also makes dye molecules degrade and lowers the COD content (Wali, 2015). The previous studies described the creation of a heterogeneous Fenton process using nanoscale zerovalent iron particles (nZVI) as a heterogeneous catalyst to treat synthetic tannery wastewater. The optimal pH was 3, the Cat/Ox (% weight) ratio was 35%, and COD removal efficiency was 68%. The initial oxidant amount (hydrogen peroxide) was fixed equal to 0.75* COD_0 (Rodriguez et al., 2019). The treatment of the common effluent (tannery effluent) and study of the pollutant removal results in terms of COD removal by both the Advanced Oxidation Process and biological treatment by a bacterial consortium were the major aims of the previous studies. With varying $FeSO_4$ doses (100–300 mg/L) and H_2O_2 dosages (400–800 mg/L), Fenton's oxidation process was required to reduce COD at different pH ranges (2–10) as a pre-treatment. The effluent was then biologically treated using bacterial (INBARAJ et al.). Under several operational conditions, such as pH, time, and Fenton doses, the effectiveness of Fenton oxidation in tannery wastewater was examined. It was noted that it produced the greatest COD removal. The COD reduction was 82% with MAC800 (mesoporous activated carbon) and 78% without MAC800 (mesoporous activated carbon). The optimum conduction was at pH = 3.5, the time of reaction was 4 hours, the concentration of H_2O_2 was 26.47 mmol/L, and the concentration of $FeSO_4.7H_2O$ was 0.1 g/L. (Sekaran et al., 2013).

The chemical oxidation by the Fenton reagent was more effective than the removal of phenol, aromatics, and non-biodegradable organic materials. The removal efficiency of phenol was 98%, and the removal efficiency of COD was 70%. When the initial concentration of phenol was 200 mg/L, the optimum H_2O_2 and $FeSO_4$ conditions were 1000 mg/L H_2O_2 and 500 mg/L $FeSO_4$. and 3000 mg/L H2O2 and 1500 mg/L $FeSO_4$ when the initial phenol concentration was

500 mg/(Yavuz et al., 2007). The photo-Fenton reaction is the response to treating organic pollutants in water.

Treatment of nonbiodegradable organic substances in wastewater that has undergone biotreatment and contains pesticides, the Fenton reagent enhanced by microwave electrodeless ultraviolet (MWEUV/Fenton) was suggested. (MWEUV/Fenton) The approach was suggested for the improved treatment of nonbiodegradable organic components in pesticide-containing biotreated wastewater. The MWEUV lamp was discovered to be more efficient for chemicals. The MWEUV/Fenton method eliminated all three major pesticides that were present in the wastewater—dimethoate, triazophos, and malathion in just under 120 minutes. The optimal conditions were found at pH = 5, concentration of Fe^{2+} = 0.8 mmol/L, and concentration of H_2O_2 = 100 mmol/L, and the removal efficiency of COD was 79.8% (Cheng et al., 2015). Table 1 describe high reactivity of Fenton oxidation process during different type of wastewater.

Table 1. wastewater treatment using Fenton oxidation process.

No	Type of wastewater	COD before	COD removal %	optimum conditions	Reference
1	Batik Cual Wastewater	104 mg/L	76.3%	0.5 g of $FeSO_4.7H_2O$, 1 mL of (H_2O_2), the Fenton method was carried out for 24 hours	(Gusa et al., 2021)
2	cosmetic wastewater	6968 mg/L	95%	pH= 3,1 mL/l H_2O_2, and 0.75 g/L for Fe(II) and Fe(III) and reaction time 40 min	(Ebrahiem et al., 2017)
3	Dyeing Bath Wastewater	--	75.81%, 78.03% and 78.14% for Disperse Yellow 23, Disperse red 167 and Disperse Blue 2BLN respectively	9ml H_2O_2, 0.12 g/L $FeSO_4.7H_2O$, pH=3 and retention time of about 100 minutes	(Wali, 2015)
4	Launderette wastewater	704 mg/L	63%	Fe^{3+} = 0.04g /L , H_2O_2 = 1.3m L and pH= 3 after 15 min	(Tony et al., 2016)

5	Tannery Wastewater	1200 mg/L	70%	(Cat/OX=35%, pH=3, the H_2O_2 amount was fixed equal to 0.75*COD_0 = 0.9 g/L (Fe^{2+}=0.18 - 0.315g/L)	(Rodriguez et al., 2019).
6	Tannery Wastewater	960 mg/L	82% of COD removal in 6% saline tannery effluent	Varying $FeSO_4$ doses (0.1-0.3 g/L). H_2O_2 doses (0.4-0.8 g/L), pH=4	(INBARAJ et al.)
7	tannery wastewater	4000 mg/L	71%	H_2O_2:$FeSO_47H_2O$ for the maximum COD reduction of tannery wastewater was 3.75g/L:0.1 g/L. The reaction time for batch study was 4 h and the pH was 3.5.	(Sekaran et al., 2013).
8	synthetic wastewater	480 mg/L 1200 mg/L	70%	1g/L H_2O_2 and 0.5g/L $FeSO_4$ for initial phenol concentration of 200 mg/L and 3g/L H_2O_2 and 1.5g/L $FeSO_4$ for the initial phenol concentration of 500mg/L. Optimum H_2O_2 and $FeSO_4$ ratio [Fe^{2+}]/ [H_2O_2] (mol/mol) equal to 0.11. pH value from 2.3 to 2.8	(Yavuz et al., 2007).
9	Pesticide-Containing Wastewater	1540 mg/L	--	The optimal conditions for MWEUV/Fenton method were at pH = 5, dose of Fe^{2+} =0.2g/L, and dose of H_2O_2 14.2g/L within 2h.	(Cheng et al., 2015).
10	Textile wastewater	202mg/L	68%		(Ribeiro et al., 2017)
11	landfill leachate	5000 mg/L	65%	10g/L of ferrous Ion. H_2O_2 = 48g/L.	(Zhang et al., 2006)
12	hospital wastewater	1350 mg/L	77%	Fe (II) = 0.135 g/L , pH =3, H_2O_2 = 5.4 g/L	(Kajitvichyanukul and Suntronvipart, 2006)
13	textile wastewaters	--	50%	0.25g H_2O_2 or 200 ml perhydrol per liter of wastewater and 0.04 and 0.05g Fe^{2+} per liter of wastewater	(Papadopoulos et al., 2007)
14	Palm Oil Effluent	6,700 mg/L	48.2%	0.8 g/L $FeSO_4.7H_2O$, pH increased from 5.3 to 7.4 within 2 h.	(Babu et al., 2010)
15	tannery wastewater	2810 mg/L	73%	pH= 3, 0.84 g/L H_2O_2	
16	tannery wastewater	6300 mg/L	72%	pH = 3.5 ratios of H_2O_2/$FeSO_4$=600/1000(w/w)	(Lofrano et al., 2010)

17	synthetic wastewater	42000 mg/L	85%	pH = 2.5, H_2O_2 concentration of 1.9 g/L	(Chakinala et al., 2008)
18	synthetic wastewater	3242 mg/L	96.3%	pH = 4, 2.5 g/L $FeSO_4$.7H2O and 100 mL/L of 30% H_2O_2, and time of 90 min	(Li et al., 2009)
19	Industrial triazophos wastewater	3418 mg/L	85.4%.	pH = 4, dosage of 5.0 g/L $FeSO_4$.7H2O and 75 mL/L of 30% H_2O_2, and time of 90 min.	(Li et al., 2009)
20	Pharmaceutical wastewater	11,987 mg/L	66.73%	Reaction time = 1.5 h; COD/H2O2 = 1:2.2; pH 3; Fe2+/H2O2 = 1:50	(Badawy et al., 2009)

4. Conclusions

Previous studies have proven that the Fenton reaction has been successfully used in the removal of many hazardous pollutants from wastewater. Where are the removal efficiencies of COD between 50% and 96%, BOD between 35% and 75%, and TOC between 40% and 70%. The treatment was effective at pH ranges between 3 and 5, and at different doses of H_2O_2 and Fe^{2+}. The amount of peroxide used and the iron concentration are two important variables that influence how quickly the Fenton reaction occurs. The importance of the peroxide dosage in attempts to increase degradation efficiency The reaction kinetics depend on the presence of iron.

References

ABDELLATIF, G., MAHMOUD, A. S., PETERS, R. W. & MOSTAFA, M. K. Waste Plastics and Microplastics in Africa: Negative Impacts and Opportunities. 2021 AIChE Annual Meeting, 2021. AIChE.

BABU, B. R., MEERA, K. S., VENKATESAN, P. & SUNANDHA, D. 2010. Removal of fatty acids from palm oil effluent by combined electro-fenton and biological oxidation process. *Water, Air, & Soil Pollution,* 211, 203-210.

BADAWY, M. I., WAHAAB, R. A. & EL-KALLINY, A. 2009. Fenton-biological treatment processes for the removal of some pharmaceuticals from industrial wastewater. *Journal of hazardous materials,* 167, 567-574.

CHAKINALA, A. G., GOGATE, P. R., BURGESS, A. E. & BREMNER, D. H. 2008. Treatment of industrial wastewater effluents using hydrodynamic cavitation and the advanced Fenton process. *Ultrasonics sonochemistry,* 15, 49-54.

CHENG, G., LIN, J., LU, J., ZHAO, X., CAI, Z. & FU, J. 2015. Advanced treatment of pesticide-containing wastewater using Fenton reagent enhanced by microwave electrodeless ultraviolet. *BioMed research international,* 2015.

EBRAHIEM, E. E., AL-MAGHRABI, M. N. & MOBARKI, A. R. 2017. Removal of organic pollutants from industrial wastewater by applying photo-Fenton oxidation technology. *Arabian Journal of Chemistry,* 10, S1674-S1679.

EL-NAAS, M. H., AL-ZUHAIR, S. & ALHAIJA, M. A. 2010. Removal of phenol from petroleum refinery wastewater through adsorption on date-pit activated carbon. *Chemical Engineering Journal,* 162, 997-1005.

EL-SHAFEI, M., MAHMOUD, A., MOSTAFA, M. & PETERS, R. Effects of entrapped nZVI in alginate polymer on BTEX removal. AIChE Annual Meeting. San Francisco, CA, 2016. 13-18.

FARAG, R., EL-SHAFEI, M. M., MAHMOUD, A. S., MOSTAFA, M. & PETERS, R. Green synthesis of nano iron carbide: preparation, characterization and application for removal of phosphate from aqueous solutions. 2018 AIChE Annual Meeting, 2018. AIChE.

GHALY, A., ANANTHASHANKAR, R., ALHATTAB, M. & RAMAKRISHNAN, V. 2014. Production, characterization and treatment of textile effluents: a critical review. *J Chem Eng Process Technol,* 5, 1-19.

GUSA, R. F., SARI, D. N., AFRIANI, F., SUNANDA, W. & TIANDHO, Y. 2021. Removing BOD, COD, and Decolorization of Batik Cual Wastewater Using Fenton Mechanism. *Jurnal ilmiah pendidikan fisika Al-Biruni,* 10, 139-148.

INBARAJ, S. T., VIJAYALAKSHMI, P. & MUTHURAMAN, G. INFLUENCE OF FENTON OXIDATION AND BIODEGRADATION ON TOXICITY OF TANNERY INDUSTRIAL EFFLUENT. *Magnesium (mg/L),* 200, 240.

KAJITVICHYANUKUL, P. & SUNTRONVIPART, N. 2006. Evaluation of biodegradability and oxidation degree of hospital wastewater using photo-Fenton process as the pretreatment method. *Journal of Hazardous Materials,* 138, 384-391.

LI, R., YANG, C., CHEN, H., ZENG, G., YU, G. & GUO, J. 2009. Removal of triazophos pesticide from wastewater with Fenton reagent. *Journal of Hazardous Materials,* 167, 1028-1032.

LOFRANO, G., MERIC, S., INGLESE, M., NIKOLAU, A. & BELGIORNO, V. 2010. Fenton oxidation treatment of tannery wastewater and tanning agents: synthetic tannin and nonylphenol ethoxylate based degreasing agent. *Desalination and Water Treatment,* 23, 173-180.

MAHMOUD, A., EL-TAYIEB, M., AHMED, N. & MOSTAFA, A. 2018a. Algorithms and statistics for municipal wastewater treatment using nano zero valent iron (nZVI). *J Environ Biotechnol Res,* 7, 30-44.

MAHMOUD, A. S., FARAG, R. S. & ELSHFAI, M. M. 2020. Reduction of organic matter from municipal wastewater at low cost using green synthesis nano iron extracted from black tea: Artificial intelligence with regression analysis. *Egyptian Journal of Petroleum,* 29, 9-20.

MAHMOUD, A. S., FARAG, R. S., ELSHFAI, M. M., MOHAMED, L. A. & RAGHEB, S. M. 2019. Nano zero-valent aluminum (nZVAl) preparation, characterization, and application for the removal of soluble organic matter with artificial intelligence, isotherm study, and kinetic analysis. *Air, Soil and Water Research,* 12, 1178622119878707.

MAHMOUD, A. S., MOHAMED, N. Y., MOSTAFA, M. K. & MAHMOUD, M. S. 2021a. Effective chromium adsorption from aqueous solutions and tannery wastewater using bimetallic Fe/Cu nanoparticles: response surface methodology and artificial neural network. *Air, Soil and Water Research,* 14, 11786221211028162.

MAHMOUD, A. S., MOSTAFA, M. K. & ABDEL-GAWAD, S. A. 2018b. Artificial intelligence for the removal of benzene, toluene, ethyl benzene and xylene (BTEX) from aqueous solutions using iron nanoparticles. *Water Supply,* 18, 1650-1663.

MAHMOUD, A. S., SARYEL-DEEN, R. A., MOSTAFA, M. K. & PETERS, R. W. Artificial intelligence for organochlorine pesticides removal from aqueous solutions using entrapped nZVI in alginate biopolymer. Annual AIChE Meeting. Minneapolis, MN, October, 2017.

MAHMOUD, M. & MAHMOUD, A. S. 2021. Wastewater treatment using nano bimetallic iron/copper, adsorption isotherm, kinetic studies, and artificial intelligence neural networks. *Emergent Materials,* 4, 1455-1463.

MAHMOUD, M., MAHMOUD, A. S., EL-SAID, M. A. & MOSTAFA, M. K. 2021b. Comparison of aluminum and iron nanoparticles for chromium removal from aqueous solutions and tannery wastewater, empirical modeling and prediction. *Emergent Materials*, 1-16.

MATAVOS-ARAMYAN, S. & MOUSSAVI, M. 2017. Advances in Fenton and Fenton based oxidation processes for industrial effluent contaminants control-a review. *Int. J. Environ. Sci. Nat. Resour,* 2, 1-18.

MOHAMED, A. S. M., BARAKA, A. M. M. & EL-SHAFAI, M. M. 2016. *Study the removal of most hazardous pollutants from aqueous solutions using nanoparticles.* Cairo University.

MOSTAFA, M. K., MAHMOUD, A. S., SARYEL-DEEN, R. A. & PETERS, R. W. Application of entrapped nano zero valent iron into cellulose acetate membranes for domestic wastewater treatment. Environmental aspects, applications and implications of nanomaterials and nanotechnology 2017–Topical conference at the 2017 AIChE annual meeting, 2017. 27-34.

PAPADOPOULOS, A., FATTA, D. & LOIZIDOU, M. 2007. Development and optimization of dark Fenton oxidation for the treatment of textile wastewaters with high organic load. *Journal of Hazardous Materials,* 146, 558-563.

RIBEIRO, M. C. M., STARLING, M. C. V., LEÃO, M. M. D. & DE AMORIM, C. C. 2017. Textile wastewater reuse after additional treatment by Fenton's reagent. *Environmental Science and Pollution Research,* 24, 6165-6175.

RODRIGUEZ, J., OCHANDO-PULIDO, J. M. & MARTINEZ-FEREZ, A. 2019. The effect of pH in tannery wastewater by Fenton vs. heterogeneous Fenton process. *Chemical Engineering Transactions,* 73, 205-210.

SARYEL-DEEN, R. A., MAHMOUD, A. S., MAHMOUD, M., MOSTAFA, M. K. & PETERS, R. W. Adsorption and kinetic studies of using entrapped sewage sludge ash in the removal of chemical oxygen demand from domestic wastewater, with artificial intelligence approach. 2017 Annual AIChE Meeting, 2017.

SEKARAN, G., KARTHIKEYAN, S., EVVIE, C., BOOPATHY, R. & MAHARAJA, P. 2013. Oxidation of refractory organics by heterogeneous Fenton to reduce organic load in tannery wastewater. *Clean Technologies and Environmental Policy,* 15, 245-253.

TONY, M. A., PARKER, H. L. & CLARK, J. H. 2016. Treatment of laundrette wastewater using Starbon and Fenton's reagent. *Journal of Environmental Science and Health, Part A,* 51, 974-979.

WALI, F. 2015. Color removal and COD reduction of dyeing bath wastewater by Fenton reaction. *Int. J. Waste Resour,* 5, 171.

WANG, J.-H., MAHMOUD, M. S. & MAHMOUD, A. S. 2022. Integrated Efficiency of Using Nanocellulose-Nano Zero Valent Iron Composite in Water Treatment. *BioResources,* 17.

YAVUZ, Y., SAVAS KOPARAL, A. & BAKIR ÖGÜTVEREN, Ü. 2007. Phenol removal through chemical oxidation using Fenton reagent. *Chemical Engineering & Technology: Industrial Chemistry-Plant Equipment-Process Engineering-Biotechnology,* 30, 583-586.

ZHANG, H., ZHANG, D. & ZHOU, J. 2006. Removal of COD from landfill leachate by electro-Fenton method. *Journal of hazardous materials,* 135, 106-111.

392d Removal of Arsenic and NORM from Water with Novel Clay Adsorbents Developed Using Camd

Ubadire Onyemaobi, Urmila Diwekar, Rajib Mukherjee, Narendra Boppana

Abstract

Arsenic, a carcinogenic metal is a widely spread contaminant in groundwater. Similarly, naturally occurring radioactive materials (NORM) including barium, strontium and radium are hazardous metals present in produced water. In this study clay-based adsorbents are used to remove the heavy metal arsenic and NORMS from water and produced water, respectively. Current separating agents of heavy metals and NORM have a limited adsorption capacity [1,2]. In our previous work, we have developed a computer-aided molecular design (CAMD) framework to design novel adsorbent candidates with enhanced adsorption capacities based on the group interaction parameters for arsenic and NORM remediation using clay materials [3]. The metaheuristic efficient ant colony optimization (EACO) algorithm is used to maximize the adsorption capacity with structural possibilities, thermodynamic property, and process conditions as constraints. The objective of this work is to synthesized designed adsorbents by co-precipitation method. The morphology and structure of the fabricated molecules were characterized using X-ray diffraction and X-ray fluorescence spectroscopy. With the help of a batch reactor at a controlled temperature, equilibrium adsorption for arsenic and NORM was performed as function of arsenic concentration in the solution, contact time and adsorbent dosage. Langmuir and Freundlich isotherms models are predicted for both arsenic and NORM. The cations present in the water is analyzed with the help of inductive coupled plasma optical emission spectroscopy. The adsorption capacity for arsenic from water and NORM from produced water is found to be an order of magnitude higher than that can be obtained by conventional clay adsorbents.

Keywords: Adsorbents, NORM, CAMD, EACO algorithm.

References:

- Benavides, P. and Diwekar, U., 2015. Optimal design of adsorbents for NORM removal from produced water in natural gas fracking. Part 1: Group contribution method for adsorption. *Chemical Engineering Science*, 137, pp.964-976.
- Benavides, P. and Diwekar, U., 2015. Optimal design of adsorbents for NORM removal from produced water in natural gas fracking. Part 1: Group contribution method for adsorption. *Chemical Engineering Science*, 137, pp.964-976.
- Doshi, R., Mukherjee, R. and Diwekar, U., 2018. Application of Adsorbate Solid Solution Theory To Design Novel Adsorbents for Arsenic Removal Using CAMD. *ACS Sustainable Chemistry andamp;amp; Engineering*, 6(2), pp.2603-2611.

392e Understanding the Mechanism of Adsorptive Selectivity of Arsenic over Phosphorous Oxoanions By Fe(III)-Crosslinked Chitosan Using DFT

Obinna Nwokonkwo, Christopher Muhich

Adsorption-based water remediation is a promising "under-the-sink" technology for removing various contaminants from drinking water. Unfortunately, the co-occurrence of target contaminants and their less-toxic, iso-structural counterparts creates significant competition for the often-limited binding sites on conventional adsorbents, thus degrading removal efficiencies. In this work, we performed Density Functional Theory (DFT) calculations to understand the key factors that control the selectivity of arsenic oxoanions, $H_2AsO_4^-$ and H_3AsO_3, and phosphorous oxoanions, $H_2PO_4^-$ and H_3PO_3, by transition-metal crosslinked chitosan. We found that pure chitosan does not preferentially bind As(V) or P(V) since there was only a marginal difference of ~0.01eV in binding energies of both species. However, crosslinking chitosan with Fe (III) facilitates coordination-complex formation with the oxoanions. Analysis of the Crystal Orbital Hamilton Populations (COHP) revealed that the selectivity of As over P is controlled by the difference in the filling of antibonding molecular orbitals in the metal-ligand bond. Differences in localization of electron density in the HOMO resulted in stronger binding of the charged species, As(V) and P(V), over As(III) and P(III). These results show that slight variations in O partial charges of the oxoanions can provide a valuable tool to distinguish between nearly identical ions and, therefore, preferentially extract one over the other.

717a Particle mechanics and solids handling in the process industries

Ahmad Haidari

Chemical and processes industries use and modify, transport, and process all types of particulates and granulate materials. It is becoming increasingly possible to bring more details about the behavior of particles at the macro-scale depending on particle characteristics (Geldart group, size, density, shape, cohesion, etc.). In a range of applications for example biomass refining, green chemicals manufacturing, and fluidized bed and catalytic reactors, understanding the flowability and behavior of discrete solid particles can lead to better understanding of design and operation of equipment and processes. In this Workshop, we will show some perspective on the role of particle mechanics and the opportunities that engineering and simulation may play in solid handling and solid-fluid equipment design.

AUTHOR INDEX

AIChE
120 Wall Street, FL 23
New York, NY 10005-4020

ISBN 978-1-7138-7902-2

9 781713 879022

CURRAN ASSOCIATES INC.
proceedings.com

Computing and Systems Technology Division 2016

Core Programming Area at the 2016 AIChE Spring Meeting and 12th Global Congress on Process Safety

Houston, Texas, USA
10-14 April 2016

AIChE
The Global Home of Chemical Engineers